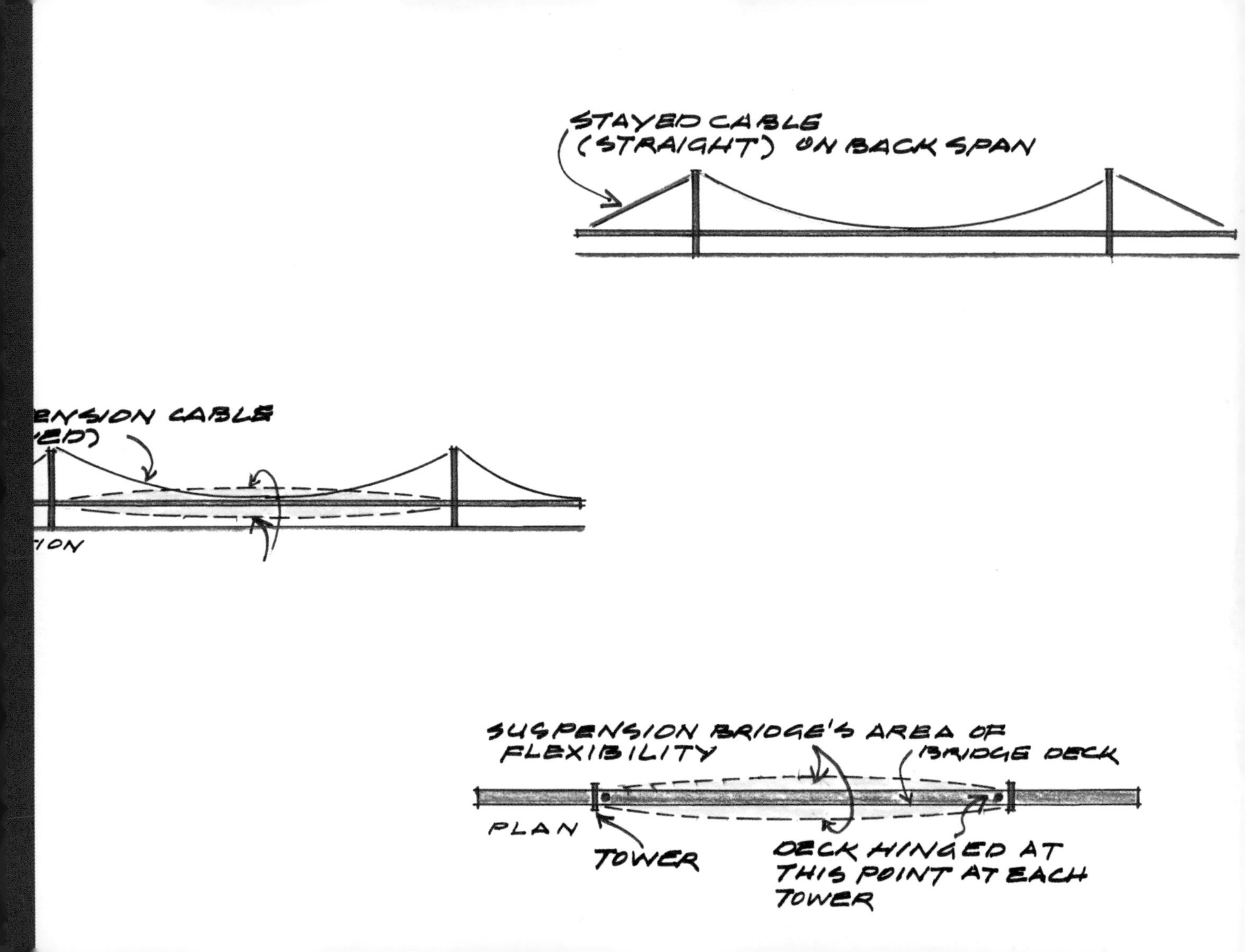
STAYED CABLE
(STRAIGHT) ON BACK SPAN
ENSION CABLE
VED)
ION
SUSPENSION BRIDGE'S AREA OF
FLEXIBILITY
BRIDGE DECK
PLAN
TOWER
DECK HINGED AT
THIS POINT AT EACH
TOWER

GOLDEN GATE BRIDGE

HISTORY AND DESIGN OF AN ICON

BY DONALD MACDONALD AND IRA NADEL • ILLUSTRATIONS BY DONALD MACDONALD

CHRONICLE BOOKS
SAN FRANCISCO

Library of Congress Cataloging-in-Publication Data available.
ISBN: 978-0-8118-6337-7

Manufactured in China.

Designed by Jennifer Lacy Bagheri.

10 9 8 7 6 5 4 3 2 1

Chronicle Books LLC
680 Second Street
San Francisco, California 94107

www.chroniclebooks.com

TABLE OF CONTENTS

INTRODUCTION
AN AMERICAN ICON

"A great city with water barriers and no bridges is like a skyscraper with no elevators. Bridges are a monument to progress."

—*Joseph Strauss,*
chief engineer, Golden Gate Bridge

The answer to the question "Why is the Golden Gate both beautiful and an architectural triumph?" is "geometry." Using a set of timeless forms, from triangles and polygons to squares and cylinders, its architects and engineers created a balanced form that appears both natural and aesthetic. As it stretches across the Golden Gate from San Francisco to Marin County, the bridge, with its suspended arc and majestic towers, dominates but does not overpower its natural setting. Its geometry uses universal forms rather than personalized architectural statements and creates a harmony of form subject to neither historical nor individual influence. The architectural integrity of the bridge, with its strongly vertical towers and the catenary, or upward, curve of its suspension cables, reflects the dynamic symmetry often found in nature.

Spanning the Golden Gate Strait—3 miles long and 1 mile wide—at the entrance to San Francisco Bay from the Pacific Ocean was an architectural as well as an engineering challenge. How could one make a beautiful bridge on such a large scale that was also useful and safe? In about 1846, the strait was named Chrysopylae, or Golden Gate, by John C. Frémont, explorer and U.S. Army topographical engineer, because it reminded him of the harbor in Istanbul named Chrysoceras, or Golden Horn. To maintain the setting's beauty and live up to its name while providing a functional bridge was the task.

The Golden Gate Bridge is unquestionably an American icon whose symbolic power rivals that of the Empire State Building or Statue of Liberty, in part through its representation of optimism, a distinctly American trait. Its visual linkage of form and nature, purpose and design, does not mark a boundary but instead symbolizes Western opportunities and Pacific possibilities. Here the American values of hope, possibility, and success mix with the future, represented by the ocean extending beyond the bridge itself.

"San Francisco, open your golden gate. You let no stranger wait outside your door."

—As sung by Jeanette MacDonald

The Golden Gate Bridge also represents the goals of the City Beautiful movement, which captured America after its showcasing at the 1893 Chicago World's Fair. The movement flourished in the early years of the twentieth century, seeking to refashion American cities and public spaces into beautiful yet functional entities. The 1901 plan for the Mall in Washington, D.C., is among its achievements. The Golden Gate Bridge, with its natural proportions and scenic location (enhanced but not altered by its design), may be the purest embodiment of its principles.

The City Beautiful movement developed urban parks, boulevards, and elegant civic centers bordered by refined public buildings separated by green space. Urban design applied to the improvement of the environment was emphasized during the movement's height (1900–1914). Scenic preservation and urban beautification were realizable goals in City Beautiful plans, which were, not surprisingly, promoted particularly in San Francisco.

Daniel H. Burnham's 1905 plan for revitalizing San Francisco expressed the new ideal. At the time the most prestigious city planner in America, Burnham sought to redesign the city's street layout and extend its park system via large vistas and new open spaces. By 1900, San Francisco ranked ninth in population among American cities but was home to only 23 percent of all Californians. It had started to lose population as early as 1880, especially to Los Angeles. Burnham's plan aimed to revivify the city physically, politically, and socially. Earlier, in 1866, Frederick Law Olmsted, landscape architect and designer of Central Park in New York, had also outlined a plan for San Francisco with a diversity of park elements united by boulevards. He also proposed a water gate on the Bay for ceremonial arrivals by sea, attached to a parade ground and pavilion for civic gatherings.

Less than a year after Burnham completed his proposal, on 18 April 1906, earthquake and fire reduced almost 4 square miles of San Francisco to rubble. But instead of turning to Burnham's plan, the city rebuilt itself on its earlier grid, preferring the secure, preexisting outline to anything new. Private property interests appeared to supersede civic ideals.

Nevertheless, the City Beautiful movement was optimistic in its belief that American cities could convert ugliness into beauty by uniting natural and classical forms. Beaux-Arts neoclassical architecture was its preferred form, as often seen in the work of Paul Cret, a French émigré whose new classicism won him the Pan American Union Building competition in Washington and the designs for the Detroit Institute of Arts, the Rodin Museum in Philadelphia, and the Folger Shakespeare Library in Washington. Cret was also one of the first architects to work successfully with engineers

Opposite: Figure 1

on bridges. He incorporated their efforts into his neoclassical projects, notably the Delaware River (Benjamin Franklin) Bridge (Figure 1) in Philadelphia and the East Boston Harbor Bridge. The Delaware River Bridge, completed in 1926, was at the time the longest span in the world and was visually distinctive for its use of masonry towers that masked the bridge anchorages, whose compressive strength contrasted with the tensile strength of steel. The concrete anchorages below the roadway at either end of the Golden Gate Bridge eliminated the need for masonry towers.

The Golden Gate Bridge is one of the last vestiges of the City Beautiful movement—this beautiful structure incorporating a panoramic landscape confirmed the movement's integration of nature, aesthetics, and urban environment. The proportion, harmony, symmetry, and scale of the bridge all exemplify many of the elements the City Beautiful movement emphasized.

In the years before the bridge was built, the City Beautiful movement was balanced by a search for an American style of architecture that combined nationalism and populism. Aiming to display America at its most beautiful and artistic, architecture and civic planning began to exemplify national ideals such as prosperity, health, and democratic choices. A quest began for an American style of architecture free from European influence, whether it was the Jugendstil school (the German Art Nouveau movement), France's Art Deco movement, or the Viennese Fin-de-Siècle style. An organic American style of architecture, as partly embodied in the work of Frank Lloyd Wright, set the tone of new public as well as private structures. This socially conscious architecture made new demands and set new goals for modernism and the International Style, the predominant, European-driven movements focused on simplified forms and industrial materials. The Golden Gate Bridge can be seen as its expression.

The Golden Gate Bridge has been described as the largest Art Deco sculpture in the world. But the romance of the bridge arises from more than its graceful,

BENJAMIN FRANKLIN BRIDGE ORIGINALLY NAMED THE DELAWARE RIVER BRIDGE · PHILADELPHIA · PENNSYLVANIA LONGEST SUSPENSION IN THE WORLD IN 1926

soaring design, which seems to ascend out of its natural surroundings, or its evocative name. Its mystique preceded even its building. At the official groundbreaking on 26 February 1933, a parade, artillery barrage, and a billowy image of the bridge painted in the sky by smoke planes greeted onlookers. Engineering students from Berkeley marched into the ceremony with an 80-foot bridge replica. Yet the bridge's history also has a practical and, at times, frightening side: its builders contended with fog so thick that workers could not see more than a few inches, treacherous crosswinds that knocked men and material into the sea, political battles that more than once threatened to halt the project, and salt and sea air that have necessitated constant repairs.

Still, from the beginning, the bridge was a national celebrity. Its numerous firsts include the first closing of the Gate to shipping, on 2 August 1935, when a barge towed cables across the Gate to form the first physical link between Marin County and San Francisco and bring California–US Highway 1 a step closer to becoming a continuous ribbon from Canada to Mexico. When U.S. President Franklin Delano Roosevelt pressed a telegraph key in Washington on 28 May 1937 to signal the bridge's official opening to the world, it set off a cacophony of church bells, foghorns, car horns, and shouts that inaugurated a week of celebrations. Thirty-eight ships of the U.S. Pacific Fleet steamed under the new crossing. Floodlights illuminated the bridge at night, turning its paint a rich gold, and it quickly became known as the "Span of Gold." On its fifty-year anniversary, in 1987, so many pedestrians walked across the bridge that the roadway flattened, returning to its arc a few hours after the deck was cleared.

Since its opening, the bridge has been thoroughly used and admired. Eighteen hundred cars and 2,100 pedestrians crossed in the first hour of its operation, and by midnight of opening day, an estimated 25,000 cars and 19,350 pedestrians had paid their tolls (fifty cents per car, five cents per pedestrian). By the end of its first fiscal year, 3,326,521 vehicles had crossed. By the end of 1970, the annual number was 33 million, and by 1996, 41 million. The all-time one-day record, set on 27 October 1989, was 164,414 vehicles. With such numbers, it is no surprise that the bridge was paid for (the cost was 35 million dollars) by 1 July 1971. And it is no surprise that in 1994, the American Society of Civil Engineers designated the bridge as one of the seven wonders of the United States, joining the Hoover Dam, the Kennedy Space Center, the Panama Canal, the Interstate Highway System, the Trans-Alaska Pipeline, and the World Trade Center—this for a bridge designed in Chicago, fabricated in Pennsylvania, and shipped through the Panama Canal. In 2001, the same society named the bridge one of eight civil engineering monuments of the millennium, and in 2007, the American Institute of Architects named it fifth among the top 150 structures in the country. (The Empire State Building was ranked first, the White House second, the Washington National Cathedral third, and the Jefferson Memorial fourth.)

The bridge's popularity and fame are balanced by its constant upkeep. Because it stands just 12 miles east of the San Andreas fault and severe weather is a continual threat, the need to upgrade and retrofit the bridge is acute. Remarkably, the bridge already has withstood high winds, storms, and earthquakes such as the devastating Loma Prieta quake of 1989. The bridge deck's ability to move 7 feet in either direction vertically or 12 feet horizontally is one secret. Another is that the bridge is composed of six elements (approaches, piers, towers, roadway, center span, and cables), not just one, so that each moves individually. But uniting all these factors is the bridge's artful merging with nature, maintaining its integrity without sacrificing beauty.

Through its combination of art and engineering, the Golden Gate Bridge has become one of the most recognizable symbols in the world. Among the signature emblems of America and a supreme work of bridge design and engineering, it is now more than seventy-one years old. Yet it never seems to age.

CHAPTER 1
ART DECO BY THE SEA

“The first consideration in designing an approach for the Golden Gate Bridge has been this concern for a modern expression.”

—*Joseph Strauss*

DESIGNING A BRIDGE

From its beginning, the Golden Gate Bridge has been synonymous with Art Deco. Its revolutionary design incorporates elements of the movement, which dominated roughly from 1910 to 1939, such as simplified geometric patterns, bold outlines, a stepped-back vertical style, and motifs like chevrons, all of which contribute to a sense of the aerodynamic. A lightness of form results, marked by the openness of the towers, through which light and space flow.

The Art Deco style demonstrated the ideal of promise fueled by prosperity. The bridge originated in this movement that redesigned everyday style, from cigarette lighters and fashion to train engines and planes. The sleekness and simplicity of its towers, its streamlined light stands, and its uncluttered bridge rail all exhibit both Art Deco’s bold outlines and powerful, repeated patterns, which reinforce the dynamic form of the structure.

Below: Figure 2
Opposite: Figure 3

THE GOLDEN GATE BEFORE THE CONSTRUCTION OF THE BRIDGE · LOOKING NORTHEAST FROM THE OCEAN (BASED ON ANSEL ADAMS 1932 PHOTO)

Art Deco triumphed in the Golden Gate Bridge design in part because of the use of architects alongside engineers. When chief engineer Joseph Strauss sought out the talents of two architects, John Eberson and Irving F. Morrow, he signaled that the architectural treatment of the bridge would be equal to its engineering elements. Until this development, engineers had acted independently, fashioning all elements of bridge design.

Such a fusion of function and form was perhaps natural for this beautiful site and its remarkable confluence of light and water, which is seen in a 1932 Ansel Adams photograph of a three-masted sailing ship heading into the open Bay (represented by Figure 2). The Bay we know today was once a wide valley, part of the system that now is composed of the valleys of Santa Clara, Sonoma, and Napa. It was originally a river gorge, until increasing water pressure deepened it enough for the ocean to invade, producing a 463-square-mile bay.

In 1775, the Spanish ship *San Carlos* was the first to sail into its uncharted narrows. A year later, the Spanish founded the Presidio and Mission Dolores (Figure 3), although the city first formed at Yerba Buena Cove under the lee of Telegraph Hill, where the waters came up against what is now Montgomery Street. Here the village prospered, becoming San Francisco in January 1847.

Early plans for bridging the crossing began ten years later when a California eccentric, Joshua Norton, proclaimed himself emperor of the United States and protector of Mexico. In a "proclamation" of August 1869, he called for a suspension bridge to join Oakland Point, Yerba Buena Island, and the Sausalito mountains. The year before, Leland Stanford had proposed a railroad across the Bay from Oakland to San Francisco, suggesting not only a walkway along its rail-bed, but also resorts and saloons to provide comfort for pedestrians. In 1872, railroad engineers for the Central Pacific proposed a less elaborate scheme, but no one came forth to build it. Instead, as early as 1910, a rail crossing over the lower Bay was built, and an extensive ferry system (represented by Figure 4) crisscrossed the Bay. Two additional plans for Bay bridges were proposed in 1915 and 1916.

MISSION DOLORES · CONSTRUCTED IN 1776 SAN FRANCISCO

Nothing happened, however, until after the First World War, when interest in a San Francisco–Oakland bridge reemerged. Thirteen companies competed in 1921 to develop the area, though the War Department rejected all proposals because of concern over the security of a naval base in the northern portion of the Bay. Still, a federal commission, established in 1929, formalized plans for the bridge,

Below: Figure 4
Opposite: Figure 5

and work began in July 1933. A northern bridge over the Golden Gate, however, still seemed a remote prospect. Indeed, geography would make it almost impossible.

No plan to cross the Golden Gate emerged until 1917, when Michael M. O'Shaughnessy, city engineer of San Francisco, asked Joseph Strauss to offer a proposal to bridge the Gate. Conveniently, this Chicago-based engineer, who lacked an engineering degree (although he took courses in the field), possessed great ambition and even greater public relations skills, both necessary to construct one of the world's most original and beautiful bridges. His expertise, however, was in building bascules, or counterbalanced drawbridges, that tilt up when opened. It was the construction of the new bascule-style Third Street Bridge (Figure 5) in 1916 in San Francisco that had first brought Strauss and O'Shaughnessy together.

The desire to bridge the Gate was fueled by the city's burgeoning population. The opening of the Panama Canal in 1915 suddenly brought major sea trade and new prosperity to San Francisco. With new residents came increased reliance on the automobile. However, the Bay's two ferry systems, one from San Francisco to Marin County across the Gate, the other to the foot of Broadway in Oakland, were becoming inefficient and outdated. Inadequate traffic facilities isolated

COUNTER BALANCE LIFT BRIDGE AT 3RD STREET
SAN FRANCISCO (ENGINEERED BY STRAUSS).

Below: Figure 6
Opposite: Figure 7

San Francisco from the northern peninsula and the eastern mainland and limited its growth.

In 1918, a survey about how to bridge the Golden Gate was made, but questions persisted. The first was how to build a link across the 5,357 feet (the narrowest part of the channel) that separated its two shores and ensure safety over the more than 300-foot depth. This would require a main span of approximately 4,000 feet. Nothing that long had ever been considered, let alone constructed. New and original construction methods had to be invented to build the structure. Fog and high winds contributed to the danger, as did ocean storms and heavy Pacific swells. Another issue was the two military reservations—the Presidio to the south and Fort Baker in Marin County to the north—that bordered the site (Figure 6). Bridge approaches would interfere with military security and buildings, and the Navy and Army objected to a bridge on the grounds that an enemy could bottle up the harbor by bombing the structure. And then there was the matter of cost: everything had to be done within a yet-to-be determined budget to forestall the political opposition of transportation interests who had long controlled the region's movement of people and automobiles. The entrance to a great harbor had never been bridged. Nature as well as man conspired, it seemed, against it.

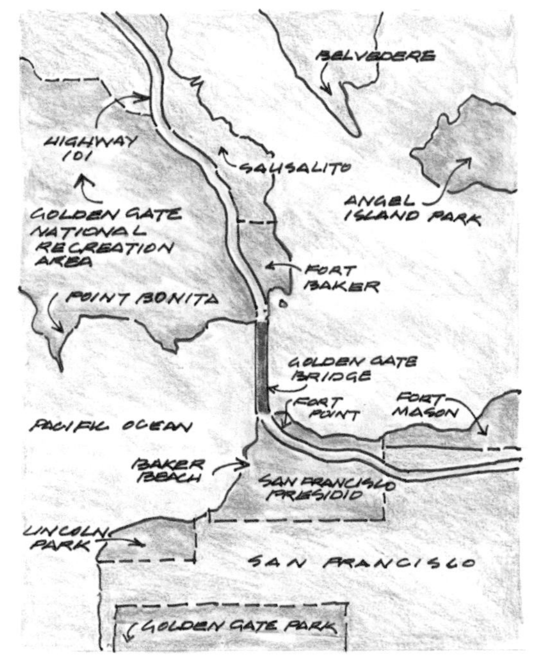

Interest, however, persisted: a topographical survey by the U.S. Coast and Geodetic Survey indicating that

THE ORIGINAL DESIGN FOR THE GOLDEN GATE BRIDGE BY THE ENGINEER JOSEPH STRAUSS · · PROPOSED 1921 ·

such a bridge could be anchored and built was completed by 1920. By 1921, Strauss had proposed preliminary drawings and an estimated budget of $17 million. The engineer was eager to attach his name to such a project and submitted a striking design to O'Shaughnessy (who did not, however, make it public until eighteen months later). Strauss outlined a symmetrical bridge with a 4,010-foot channel span constructed with 685 cantilever arms and a 2,640-foot conventional suspension span between them (Figure 7). Vertical navigational clearance was 200 feet at the center, the towers were 995 feet high, and the single deck provided space for four traffic lanes and two 7-foot-wide sidewalks. The concept combined a cantilever bridge with a suspension bridge. The problem of excessive weight was overcome by the latter, while the rigidity of the former lent the structure strength. The new Bridge District, formed to oversee development, adopted the design as its logo and initiated public meetings.

Below: Figure 8

By 1923, the Bridging the Golden Gate Association had formed, with Strauss as its chief engineer. The State of California incorporated and approved the Golden Gate Bridge and Highway District Act on 25 May 1923. "Bridge the Gate" became the association's rallying cry, especially in the northern coastal counties.

But Strauss's 1921 design was problematic. First, could one successfully combine the cantilevered and suspension bridge forms in an attractive way? Second, were the piers properly anchored? And third, how much would it cost? Although San Franciscans and residents of most surrounding counties (Figure 8) accepted the bridge proposal, its design was still unsettled. Although the public did not initially object to Strauss's design, others did, including Charles Derleth Jr., chair of Berkeley's Engineering Department. An authority on bridges, he favored an all-suspension bridge for its flexibility, lightness, and ability to cover large distances.

SUSPENDING A BRIDGE

Suspension bridges (Figure 9) have an unusual history. The first regular suspension bridge was built in America in 1796, and by 1810, fifty more had been constructed. Originally, their supporting cables were made of hand-forged chains. John A. Roebling innovated on that design by using rope woven of wire to suspend bridges over rivers, designing first a successful bridge over the Allegheny River at Pittsburgh. In 1854, he built a wire cable suspension bridge over the Niagara Falls rapids, and then, between 1870 and 1883, the Brooklyn Bridge, a distinctive structure with masonry towers and cables of woven steel wire (Figure 10). Later, the 1926 Delaware River Bridge (now called the Ben Franklin Bridge), linking Philadelphia to Camden, New Jersey, increased the length of bridges' clear, unsupported spans by using a new cellular-plate steel tower structure and stronger cables—techniques that would make construction of the Golden Gate Bridge easier. Beaux-Arts architect Paul Cret was its supervising architect and Leon Moisseiff, later advisory engineer for the Golden Gate Bridge, its design engineer. The George Washington Bridge (Figure 11), built in 1931 and engineered by O. H. Ammann with the assistance of architect Cass Gilbert, extended the method. Its new features included a suspended roadway built without stiffening trusses,

Figure 9

MENAI BRIDGE, WALES BY THOMAS TELFORD 1826.

Figure 10

BROOKLYN BRIDGE · NEW YORK · CONSTRUCTED IN 1883.

Figure 11

GEORGE WASHINGTON BRIDGE · NEW YORK · AT THE TIME OF CONSTRUCTION (1931) THIS BRIDGE HAD THE WORLD RECORD FOR THE LONGEST SPAN AND THE HIGHEST TOWERS

Figure 12

10'
CONSTRUCTED OF
WROUGHT IRON
985' HIGH
STRAUS'S
ORIGINAL
BRIDGE
DESIGN
FINAL
BRIDGE
DESIGN
TOWER
746' HIGH
EIFFEL TOWER · PARIS · FRANCE
TALLEST MANMADE CONSTRUCTION
IN THE WORLD AT TIME OF ASSEMBLY
IN 1889.

ALLAN RUSH'S SUSPENSION BRIDGE DESIGN · 1924
THIS DESIGN SHOWED, EARLY ON, THE GRACEFUL
STRUCTURE OF THE SUSPENSION BRIDGE CONCEPT.

below which a second deck could be added (as it was in 1962). Until completion of the Golden Gate Bridge, it held the record for tower height and length of span.

The cost of Strauss's 1921 combined design was within the estimated budget of 17 million dollars. Strauss himself promoted the work, pointing out to eager audiences in Marin and other northern counties that the new bridge would instantly increase property values, encourage developers, stimulate construction, and encourage tourism. A bridge was not hard to sell, especially when he noted that it would be two and a half times larger than any similar bridge and its towers ten feet higher than the Eiffel Tower (Figure 12). Not surprisingly, William Randolph Hearst's San Francisco *Examiner* strongly supported the bridge concept because he understood what it would do for the area economy. In 1924, Allan Rush, a Los Angeles engineer, proposed a more graceful, futuristic, and bold suspension bridge with taut cables (Figure 13). Though its ship and wind clearances were flawed, it appealed to many and

Opposite: Figure 13

made Strauss's proposal look antiquated. Yet Rush's proposal lacked wide public support, while Strauss, to win over the War Department, politicians, and the press, modified his design, raising the towers and adding a Parisian motif (Figure 14): the toll plaza would have an ornate entryway modeled on the Arc de Triomphe (Figure 15). In 1924, the War Department approved the bridge. And for the moment, his combined cantilever-suspension design remained.

Yet in 1926, the Joint Council of Engineering Societies of San Francisco requested a series of new engineering studies, which indicated that Strauss's original proposal had structural flaws. Suddenly, politicians and backers turned against him, and by 1929, others had applied for the job of chief engineer, including Ammann and Moisseiff, then at Columbia University and the leading theoretician of bridge design, who earlier had been a consultant for Strauss. In November 1925, Moisseiff had prepared a report on a comparative design of a suspension rather than a cantilever-suspension bridge. It outlined a structure similar to the one finally built, but would have cost 19 million dollars rather than Strauss's original estimate of 17 million. His 4,000-foot suspended span, twice the length of any yet built, seemed outrageous and was easily rejected in favor of Strauss's more secure design.

A 1927 report began to question Strauss's original plan in more detail, pointing out that the pier foundations supporting the bridge would be difficult to construct and that the stiff steel superstructure would parallel the nearby San Andreas fault (Figure 16) and would not provide appropriate lateral stability. Allowances for the dead (permanent) weight of the bridge itself were also inadequate. Further reports found geological studies incomplete, financial data inadequate, and political support eroding. New proposals were needed.

The new review did not deter Strauss, although he realized that his plan was in jeopardy, largely because the two rock shoals near the Gate's shoreline might not extend out far enough to bear the heavy towers required

STRAUSS DESIGN FOR BRIDGE ENTRY (MODELLED ON THE ARC DE TRIOMPHE)

Opposite: *Figure 14*
Below: *Figure 15*

to support his bridge, as the 1927 report detailed. If the towers had to be built on either shore, the bridge's cost would be prohibitive. Strauss had never built a suspension bridge, but his determination to build the crossing surmounted all obstacles, not least his realization that he must reject a design he had spent eight years promoting.

ENGINEERING A BRIDGE

Strauss, now with the help of structural engineer Charles Ellis, professor at the University of Illinois and vice-president of Strauss's Chicago firm, set out to redesign the original plan, whose fusion of dark functionalism and ornate wedding-cake design displeased experts. At the same time, the Bridge District went to voters for approval of a bond issue to raise money for the bridge. Strauss outflanked his opposition and managed to co-opt his former competitors Ammann and Moisseiff, who started to work with him as advisory engineers. Charles Derleth also joined the group.

Strauss was a man who had never built a bridge of this scale and whose first design could not be built within budget. But it seemed he had the job—after much politicking and lobbying, Strauss became the Bridge District's official engineer in 1929.

Through many exchanges between Ellis and Moisseiff, the reconceived design took shape. As it was refined, proposed rail lines for trolleys were eliminated in favor of six traffic lanes for cars and buses. Strauss also enlisted the aid of another engineer, Clifford Paine, an operational wizard who became his principal assistant engineer. Most of the design, detailing, plans, and computational work were done at the Strauss and Paine head office in Chicago, where at one time forty men worked on the project. During the last three years of construction, 1934–37, Paine moved to San Francisco to supervise fieldwork.

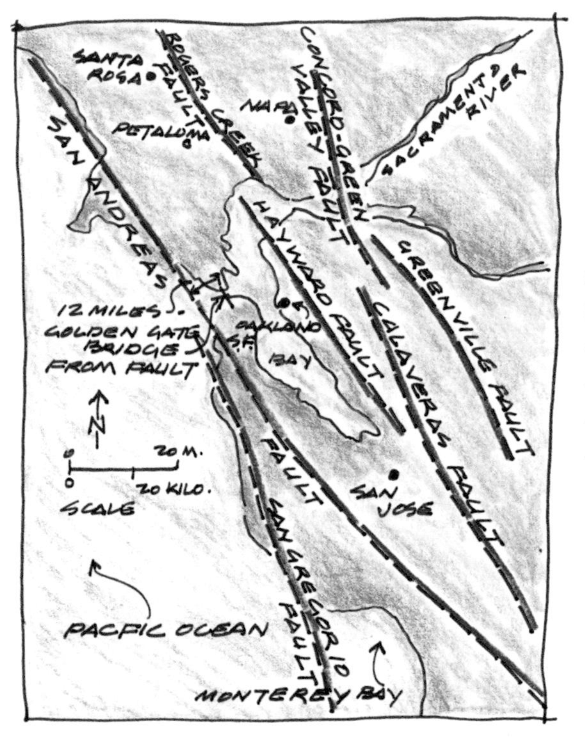

MAP SHOWING TRACES OF ACTIVE FAULTS IN SAN FRANCISCO BAY REGION

Strauss soon shifted his support to a new all-suspension bridge. His original idea, with its heavy metalwork, was too costly and time-consuming to build, and unattractive to boot.

Meanwhile, Moisseiff's calculations showed that lighter, longer, and narrower spans could withstand high winds. Strauss now recommended a suspension bridge with curved backstays—cables attached at the top of a tower, extending to and secured in an anchorage to resist stresses—rather than straight or stiff ones (Figure 17), which would require more piers on the San Francisco side, violating the original permit plans approved by

Opposite: Figure 16
Below: Figure 17

the War Department. Analysis, surveys, and test borings confirmed that a suspension bridge with two piers properly positioned to stabilize it could indeed be built. And support grew throughout 1929 for what would be at the time the longest single-span suspension bridge in the world, using the most advanced design possible.

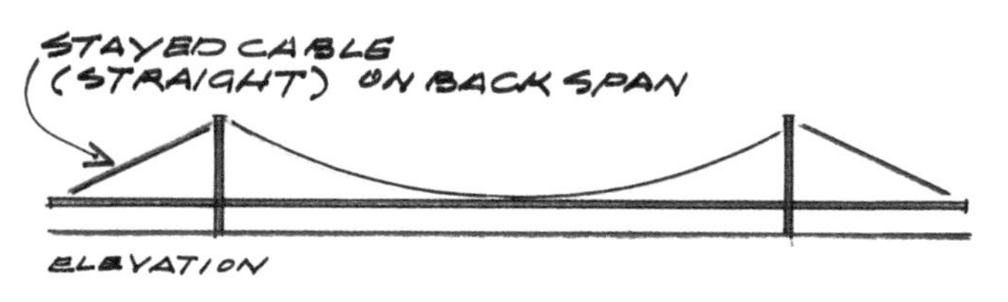

Who actually designed the Golden Gate Bridge? This is a complicated question. Strauss claimed credit for work done by others, notably Ellis and the San Francisco consulting architect, Irving F. Morrow (who replaced the first consulting architect, John Eberson). Many other players, too, were required to meet the design challenges. Bridge designs until the late 1920s often had been determined by materials and construction methods. But could one integrate structural physics with aesthetics? The Golden Gate Bridge designers successfully proved they could, removing cross trusses in the towers (Figure 18) and replacing them with horizontal struts that allowed for open portals, and placing the road deck just one-third up the towers' height, creating an aesthetically pleasing imbalance and sense of airiness—which the Bay Bridge, with its roadway halfway up its towers, lacked (Figure 19).

Strauss certainly had determination—and some would say vision—but recent evidence suggests that Ellis did the principal design work, including the stress analysis of Strauss's revised design, while proofing calculations and equations for his own modifications, which Strauss adopted. His technical knowledge of the bridge far surpassed that of Strauss, who may have inspired but did not actually engineer the structure. That was Ellis's triumph.

Figure 18

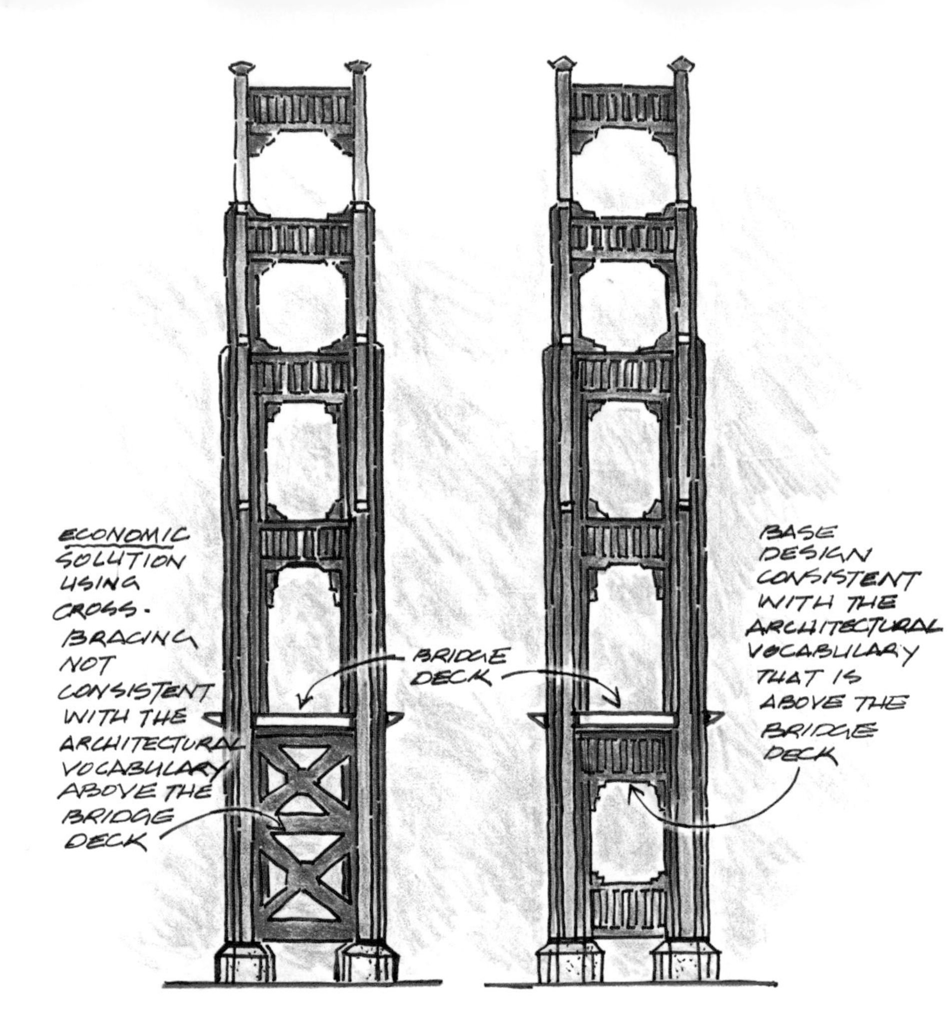
ECONOMIC SOLUTION USING CROSS-BRACING NOT CONSISTENT WITH THE ARCHITECTURAL VOCABULARY ABOVE THE BRIDGE DECK
BRIDGE DECK
BASE DESIGN CONSISTENT WITH THE ARCHITECTURAL VOCABULARY THAT IS ABOVE THE BRIDGE DECK
COMPARISON OF MAIN TOWERS WITH DIFFERENT STRUCTURAL BASES

Figure 19

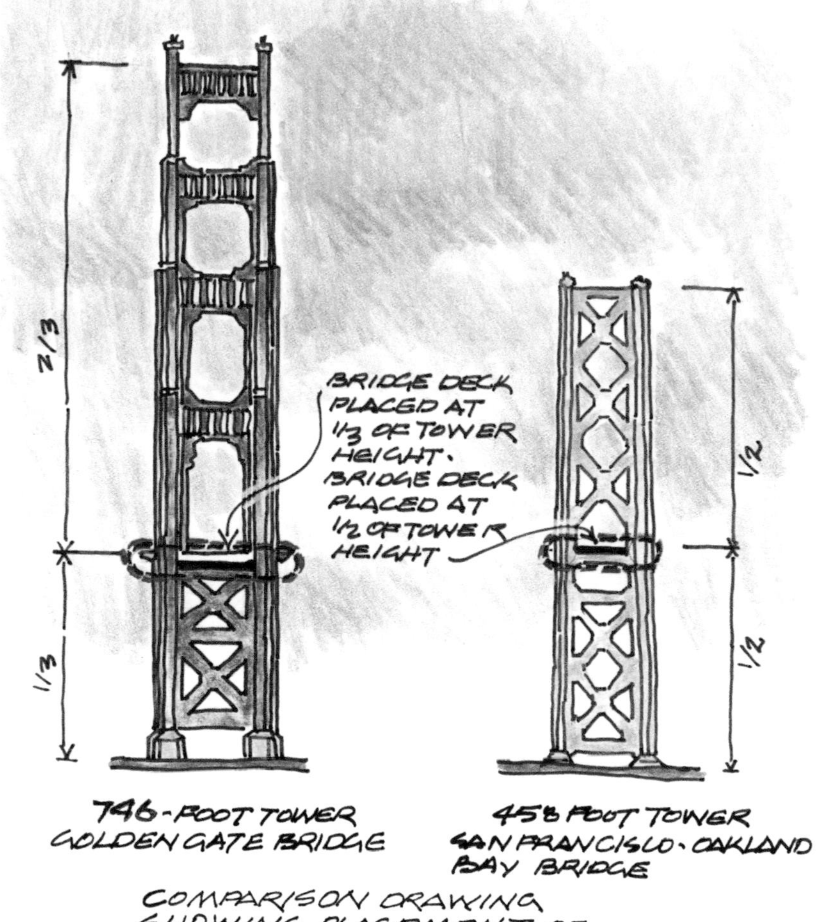
2/3
1/3
BRIDGE DECK PLACED AT 1/3 OF TOWER HEIGHT.
BRIDGE DECK PLACED AT 1/2 OF TOWER HEIGHT
1/2
1/2
746-FOOT TOWER
GOLDEN GATE BRIDGE
458 FOOT TOWER
SAN FRANCISCO-OAKLAND
BAY BRIDGE
COMPARISON DRAWING
SHOWING PLACEMENT OF
ROADWAY DECK

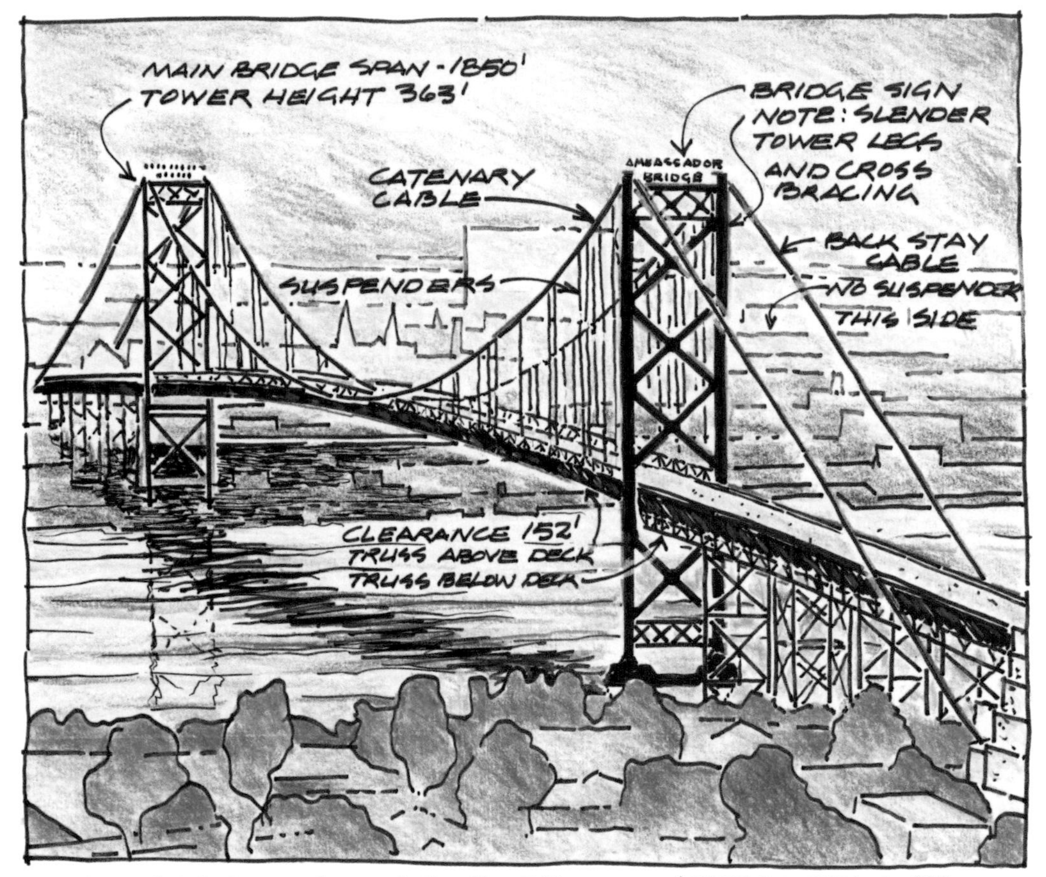

AMBASSADOR BRIDGE · DETROIT – AT THE TIME OF CONSTRUCTION (1929) IT WAS THE LONGEST SUSPENSION BRIDGE IN THE WORLD.

Opposite: Figure 20
Below: Figure 21

Ellis was also the man with day-to-day responsibilities for the bridge, while Strauss was its promoter and spokesperson. Ellis, by 1930 operating out of the Chicago office of Strauss Engineering and consulting with Moisseiff, who was now in New York, calculated stress, load factors, height, earthquake factors, and wind shifts.

STYLE AND THE BRIDGE

Consulting architect John Eberson also played a key design role. Hired in 1930, Eberson previously had specialized in atmosphere and illusion, using Art Deco's swooping lines to create elaborate Italianate and Moorish interiors in theaters and movie houses. Floating screens, marquees with changeable lettering, and the use of structural glass in exterior design were among his innovations. Although Eberson had never worked on a bridge before, Strauss believed he could contribute romance and drama to the design, especially the towers and anchorage. The towers were among the design's most critical components, its most striking feature and the largest built to that date. Advances in metallurgy allowed greater flexibility and height, which suggested lighter towers of a less mechanical appearance and the possible elimination of cross-hatch beams that could disrupt the design. The Ambassador Bridge in Detroit (built between1927 and 1929) showcased the use of such towers (Figure 20), offering open space and a light look. But could great open portals to

Figure 22

DRAWING SHOWING
THE STEPPED
TOWER LEGS
STEPPED MAYA TEMPLE
OF TIKAL · GUATEMALA.

Figures 23 & 24

ILLUSTRATION SHOWING HUGH FERRISS SKETCH OF THE IMPLICATIONS OF NEW YORK'S 1916 ZONING LAW

STORER HOUSE · LOS ANGELES · DESIGNED IN THE MAYAN STYLE BY FRANK L. WRIGHT IN 1923.

the sky, allowing in light and air, work for the proposed Golden Gate Bridge? Eberson leapt at the idea.

The towers were also distinguished by their stepped-off, or indented, form, then in vogue in such skyscrapers as New York's Chrysler Building (Figure 21). Eberson understood that indenting the towers as they rose would enhance their monumentality, height, and grace. The towers grew smaller as they grew taller, reversing conventional bridge design. The origin of this style was the Mayan stepped-back pyramid form—a precursor to the American skyscraper, with its traditional progressive setbacks—which made structures more earthquake-resistant. Without neoclassical or other stylistic camouflage, the pyramids gain height and monumentality through their setbacks, which increase as the pyramids become higher. Their scale of detail, meanwhile, actually decreases as one goes higher, which creates a sleeker and more imposing structure. Unknowingly, perhaps, Eberson borrowed this technique for the 746-foot bridge towers (Figure 22).

FRONT ENTRY AREA OF THE AZTEC HOTEL · MONROVIA CALIFORNIA

NOTE: WALL DECORATION SIMILAR TO THE WALL DETAILS FOUND IN THE MAYAN RUINS OF TULUM · MEXICO

As early as 1891, architect Louis Sullivan had suggested using setbacks to lighten bulk and increase street-level sunlight amid massed urban buildings. New York zoning laws, starting in 1916, actually required the use of setbacks (Figure 23). Many American skyscrapers soon exhibited the pyramidal form, with clean planes punctuated by

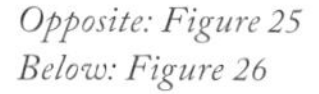
Opposite: Figure 25
Below: Figure 26

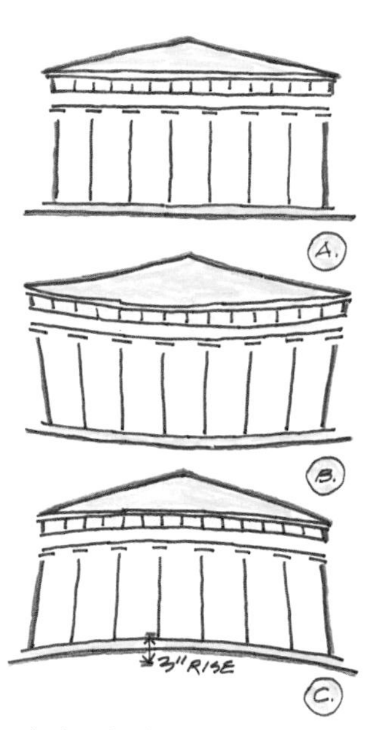

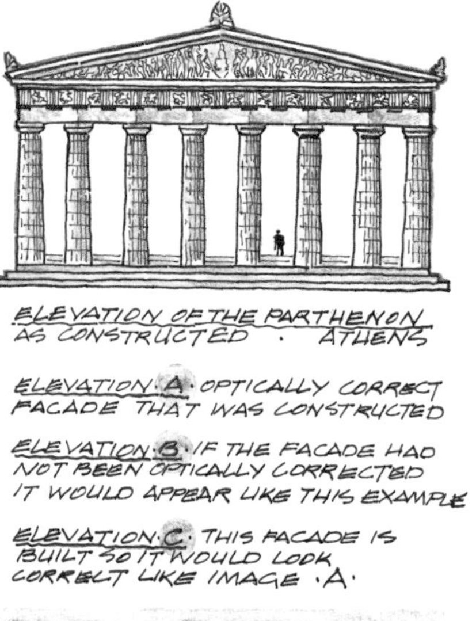

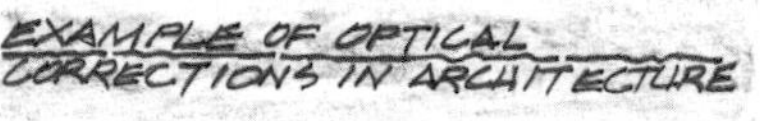

geometric lines. On the West Coast, the Mayan style began to be seen in buildings like Frank Lloyd Wright's Storer House in Los Angeles (Figure 24) and Robert Stacy-Judd's Aztec Hotel in Monrovia, California (Figure 25), whose intense relief work is echoed on a grander scale by the bridge.

Examining the towers proves the step-back's value. As they ascend, they appear thinner, which makes them look taller, precisely the principle the Greeks employed when shaping the columns of temples such as the Parthenon, which are narrower at the top than at the bottom, emphasizing their balance and enhancing their height. As they taper inward, they appear to grow taller without losing any of their beauty (Figure 26). Offsetting this effect and creating spatial tension on the Golden Gate Bridge is the elongation of each tower's four portals as it ascends. The first portal is wide and tall at its bottom to accommodate the roadway; the next is oblong; the next appears as a rounded square; and the top portal appears wider than it is high (see Figure 18, earlier). Their variation gives the towers a painterly sense of perspective. Both Ellis and Eberson contributed significantly to the design of the bridge and its towers, adding to the artistic sense of form that defines the structure.

Below: Figure 27
Opposite: Figure 28

ART DECO IN THE AIR

RUSS BUILDING · 235 · MONTGOMERY STREET SAN FRANCISCO CONSTRUCTED IN 1927

Eberson also introduced Art Deco decorative motifs to the bridge. While the bridge embodies the Americas—both in its elements of Mayan and native art and architecture, which had become an alternative to the Greek and Roman neoclassicism found on numerous public buildings, and in its expression of optimism at a time of economic discouragement—its strongest design influence was Art Deco. The style is seen in Eberson's suggestion of a proscenium at the corners of the portals and the horizontal tower bracing's vertical fluting, which accents sunlight and thereby reduces the towers' bulk and visual weight, enhances the structure's lightness, and provides depth. Art Deco motifs establish a unifying visual identity for the bridge and transform the structure into sculpture.

The Paris Exposition of 1925 formally introduced Art Deco. (The term itself originated in the official title: "L'Exposition Internationale des Arts Décoratifs et Industriels Modernes.") The United States was invited to participate but did not, believing it lacked a unique style or art, but a delegation of American architects, artists, and designers did attend—in the late 1910s and early 1920s, as advanced materials technology in products like steel allowed for the construction of skyscrapers, a modern American style was emerging. And after 1925, industrial design began to incorporate Art Deco elements, including simplified forms, planar

ENTRY DETAIL OF 450 SUTTER · BUILDING DESIGNED BY TIMOTHY PFLUEGER SAN FRANCISCO (MAYAN MOTIFS).

exteriors, strong contrasts, rigid vertical lines, and strong geometric patterns. The result was a kind of non-decorative, sleek, stylized theatricality expressed through materials like brushed aluminum and black Bakelite, stark contrasts in color and intensity, and geometric angularities.

In American architecture, Art Deco grew dominant in its emphasis on objects such as cubes, planes, and other geometric forms. It stressed the contrast between lights and darks while underscoring bold, simplified shapes, with an emphasis on mass and volume. It used streamlined forms, parabolic curves, and smooth, sleek finishes that visually embodied so-called speed lines. As its design vocabulary expanded, Art Deco was seen, too, in the graphic arts, with bolder but still streamlined lettering and shapes. Complementing Art Deco was Cubism, with its favored geometric motifs—zigzags, chevrons, rays, stepped arches—and stylized floral and natural forms. Simplified detail, such as incised lines or reeding, replaced excess ornamentation; vertical fluting, consisting of intersecting plane facets that could be readily carried out in concrete and steel, gives the Golden Gate Bridge, in particular, remarkable continuity while enhancing its monumental verticality.

COIT TOWER ON TELEGRAPH HILL IN SAN FRANCISCO · BUILT IN 1934

Opposite: Figure 29

Its hand railings, light posts, and approach structures all repeat the strong vertical lines.

The movies were perhaps Art Deco's most important promoters. Glamorously dressed actors walked or danced in sleek sets showcasing the latest in decorative style. Everything was streamlined in film: nightclubs; dance palaces; apartments; the dress, silhouette, and behavior of the actors and actresses. Even smoking became an Art Deco form, evoking terms like *suave*, *sophisticated*, and *debonair*. The cocktail became the favored drink, served in glasses that embodied Art Deco, their thin stems leading to a wide, stylized lip that spread outward like a flower.

In San Francisco, Art Deco was everywhere, taking such notable forms such as the Russ Building at 235 Montgomery Street (Figure 27), a large structure lightened by its stepped profile; the arch with elongated molded-plaster light forms surrounding the door at 450 Sutter Street (Figure 28); and the well-known elongated and fluted Coit Tower (Figure 29), an early symbol of the city. The blue-green terra-cotta of the I. Magnin building at 2001–11 Broadway in Oakland is a stunning example of San Francisco Bay Art Deco, as is the Paramount Theatre at 2025 Broadway in Oakland (Figures 30–31), designed by Timothy Pflueger, the area's leading Art Deco architect. In the late '20s and early '30s, San Francisco was replete with examples, so it is not unusual that the bridge would incorporate Art Deco elements. They were part of the Bay Area's design vernacular.

Although the ideas of John Eberson, principal translator of Art Deco to the bridge, were costly, they were adopted into its design. Yet at the same time, Strauss learned that because of road realignment mandated by federal authorities, he would need a completely redesigned bridge plaza. To complete the job and on the recommendation of California painter Maynard Dixon, in 1930 he replaced Eberson with local architect Irving F. Morrow, who lacked Eberson's national reputation but knew local politics, landscape, and art,

ART DECO · PARAMOUNT THEATER
OAKLAND · CALIFORNIA

Opposite: Figure 30
Below: Figure 31

DETAIL OF THE ART DECO TILED MURAL ON THE FACADE OF THE PARAMOUNT THEATER

and maintained an interest in painting throughout his career. He lived in Oakland and daily crossed the Bay by ferry, growing aware of the constant interplay of light and weather on the Golden Gate. Morrow believed that American architecture should stand on its own, reflecting national, not European or classical, design. American Art Deco might be the way. Working with assistant engineer Clifford Paine, who believed the bridge should be undecorated, the two refashioned the bridge's design and engineering.

Although principally concerned with the entryways and plazas, Morrow also improved the towers, making their portals rectangular and of similar shape but varying size to produce a ladder-like rise. Each tower now seemed to be a stack of slightly smaller but connected towers. An idea that Charles Ellis proposed was to drop the giant cross-members and the cross-beams of the towers below the roadway but above the pier to make the towers seem lighter and more graceful, eliminating the Erector Set look that plagued many bridges, including the Bay Bridge. But cost scrapped these plans.

"Make it beautiful," Strauss instructed Morrow in a letter. Morrow's interest in graphic arts, the most popular Art Deco medium, allowed him to see the bridge as sculpture, albeit a sleek and modern one. By 1929–30, the design had evolved into one that visually eliminated the mechanical in favor of a sleek balance of lightweight towers and artfully curved suspension cables, which today causes one to think of the bridge as a giant harp. Construction was about to start.

CHAPTER 2

COLOR, OR WHY ISN'T THE GOLDEN GATE GOLD?

"Throughout a considerable portion of the year, high fogs render the light of San Francisco colorless or gray."

—*Irving F. Morrow, bridge architect*

Why isn't the Golden Gate Bridge gold? Contradicting its name, the bridge is currently painted in International Orange, a color chosen after a long and sometimes unintentionally comic debate. The color has always been controversial; many early engineers and architects concluded that the bridge should be gold—especially because local rain and light create grayness in fog, cool blues in sunlight. Initially, the Air Force wanted the bridge striped in orange and white; the Navy wanted yellow and black. Irving Morrow wanted to paint its pylons in shades of gold (Figure 32), while some thought aluminum would be more impressive. The first coats of paint applied were actually red. Even in later years, the debate lingered: San Francisco poet laureate Lawrence Ferlinghetti included painting the bridge golden in a "wish list" at his 1998 inauguration. Yet today the International Orange bridge strikingly contrasts with the blues and grays of the water and sky, echoing the color of the iron-rich soil in the Marin Headlands at its north end and glowing at sunrise and sunset.

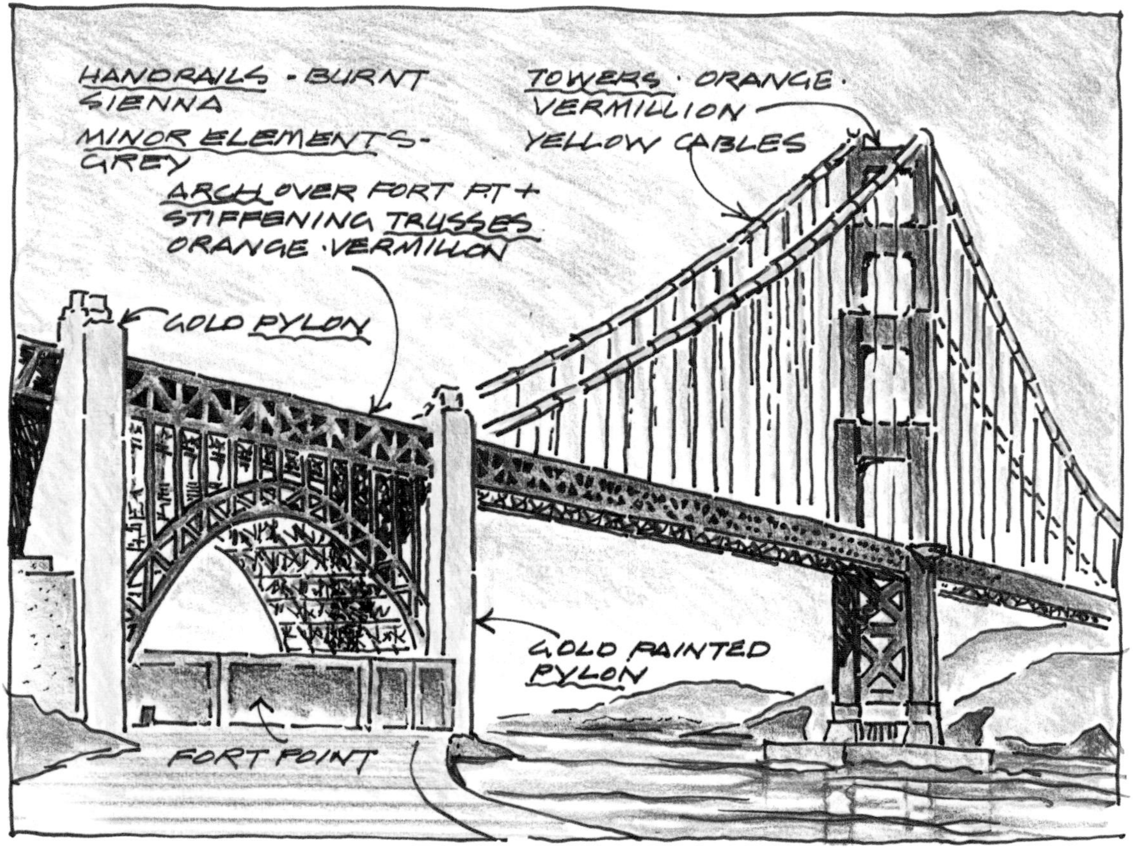

HANDRAILS - BURNT SIENNA

MINOR ELEMENTS - GREY

ARCH OVER FORT PT + STIFFENING TRUSSES ORANGE · VERMILLON

TOWERS · ORANGE · VERMILLION

YELLOW CABLES

GOLD PYLON

GOLD PAINTED PYLON

FORT POINT

COLOR SCHEME SHOWING ONE OF MORROW'S COLOR SYSTEMS WITH YELLOW CABLES AND GOLD PYLONS

Opposite: Figure 32

In the early years of the bridge's design, pale aluminum, funereal black, dull gray, and garish orange stripes were all proposed and rejected alternatives. Aluminum was too fragile a color, it would appear to thin massive weight-bearing structures like the bridge towers. Black would reduce the bridge's scale, suppress its modest ornamentation, eliminate visual variation, contradict its grace, and constantly conflict with local landscape colors. Warm grays might have been preferable, but gray offers no particular distinction. The prevalence of fog would also make a gray bridge dangerously invisible. Given cool overall atmospheric colors and what the press called the "timidly color-less" buildings of San Francisco, it was decided that the bridge should be of contrasting or warm colors. Warm golds or reds are naturally absent in the local landscape and architecture; consequently, yellow, orange, and red were considered. But yellow would lack substance; deep reds would be too heavy and lack luminosity. As one report stated, a poorly chosen color could create disharmony between the structure and its site. The bridge's expressive design and imposing structure had to be balanced by the right color, which would enhance its form and set off its large scale.

And the ideal color was already evident: the primer coat of pure red lead applied as the towers were constructed. It was luminous, responded magnificently to changes in atmospheric light, and was prominent without being insistent. Best of all, it gave weight and substance yet was light enough to register variations of shade and shadow. Thus, when Morrow made his first recommendation to the bridge's Engineering Board on 16 July 1934, he suggested orange-red for the towers, with deeper shades for the suspenders, cables, and approaches. Though he agreed that a bright color was desirable, engineer O. H. Ammann objected. In his work on the George Washington Bridge, he had had no luck in finding durable red paint. In response, a manufacturer's representative assured the board that such a paint could be created. A board member then asked if the towers' portal enclosures should be painted to foreground their modern design. No, said Morrow—any treatment in a different color would divide up the towers in a way fatal to their scale.

Morrow's friendship with *en plein air* California painter Maynard Dixon convinced him that International Orange was the right choice. In 1935, Dixon wrote a letter about the color debate to the San Francisco *Chronicle*: "Imagine Los Angeles overlooking such a chance! If these bridges [the Bay Bridge and Golden Gate Bridge] are truly glorious achievements, then as truly they must LOOK glorious! . . . By all odds, let us have them painted in red lead or orange mineral. We need it for ourselves and for our visitors." He advised that graphite (black; Figure 33) and silver (aluminum) be avoided; color was what the bridge demanded, but it must be the right one. Maynard, eventually consultant on the bridge, further noted that red lead primer's impermanence was a plus: its fading would harmonize with atmospheric effects, and irregular variations in tone due to repaintings would enhance the bridge's picturesqueness. Not all agreed. One reader responded that primer was associated with construction work and, therefore, with unfinished structures. Others felt that the favored orange-vermilion color would be less effective than light gray (Figure 34) or aluminum when lit at night. But Morrow thought that color was more significant than illumination.

The Air Force had yet another idea: in a 2 April 1936 letter, it proposed orange and white stripes (Figure 35). In fact, 63 percent of all U.S. transport pilots favored striping the towers to make them visible to aircraft. The author of the Air Force letter noted that the aluminum color of the Bay Bridge made it a hazard to flying operations. Why? The aluminum blended in with the gray fog, making the towers and structure virtually invisible. The Navy, meanwhile, proposed yellow and black stripes (Figure 36).

As color was debated, other considerations included the bridge's shadows and modeling (its chevrons and fluting). Were a dark color chosen, the former would be invisible and the latter disappear. The structure would lose its texture—only its silhouette would be noticeable—and its apparent size would be reduced. Luminous colors, by contrast, would increase scale. Painting the bridge in a single tone, a 1935 report emphasized, would produce a monotonous effect because of its unprecedented scale; a variety of tone was necessary. The report also suggested that the towers use the same basic tone throughout their height, but that a slightly darker hue be applied to diagonal bracing below the deck. A similar color would be used on the arch over Fort Point. An even darker color would be used for the San Francisco approach viaduct, the Marin approach, and the cables. The hand rail would be of a contrasting color, while miscellaneous minor details would be painted a neutral gray. The different but graduated tones would all be unified, however, and focus interest at the structure's most important point: its center.

All color treatment, Morrow urged in a report, should avoid cleverness and aim at "the utmost breadth and simplicity of effect." One reason for this is that, unlike other bridges, the Golden Gate Bridge is placed in a significant geographic location. Most bridges seem to be casually located. This one's position, however, appears inevitable, giving it an almost formal, architectural relationship with its setting that demands a conspicuous color treatment.

In the end, International Orange was selected because it echoed the Marin headlands and contrasted with the gray and blue sky and water. Technically known as basic lead chromate (also as International Airways Orange, because it is a standard for marking runways), the paint dramatically enhances the bridge's visual structure. Paint made from basic lead chromate also retains its color over a long period of time under all conditions and dries to a tough, protective elastic coating that does not crack or scale. The red, earthy color met all practical and aesthetic criteria: for some it

Figures 33 & 34

Figures 35 & 36

COLOR SCHEME BY THE U.S. AIRFORCE

COLOR SCHEME BY THE U.S. NAVY

recalled the colorful sails of Italian fishing boats passing through the Gate, their rich hues of rust red and red-orange combining with the tones of the Bay and hillsides. For others, it echoed the color of the old International Harvester logo. Soon the Golden Gate Bridge earned the sobriquet "the Red Bridge."

International Orange is not the only color on the finished bridge, however. Looking closely, one can see gray on the "oddball" or service elements of the structure —ladders, piping, electrical transformers, rail guards, and cleaning equipment—which causes them to drop from view, blending in with the sky or weather and adding to the visual and artistic integrity of the overall form.

Salt air and fog necessitate the repainting of the entire bridge every four years or so, an activity that quickly became the most costly maintenance item in the Bridge District's budget. Painting actually began in 1934, once the steel towers began their skyward ascent. Initially, men dangling on precarious scaffolds or in bosun's chairs high above shoreline rocks or waves applied a primer coat to protect the structure from corrosive salt air. Today, thirty-eight men make up the painting crew, working in freezing winds, fog, and unpredictable and changing conditions. The men prefer to work on the Marin tower and the north end of the span because the climate there is more consistent, drier, and calmer. Winter rains and summer fogs always make the work treacherous, however, as does wind—it never stops blowing, but only changes velocity. The weather is so violent and disruptive that some ninety days of work a year are lost.

As the painters in their rigging battle the sea air and fog, they are assisted by a group of seventeen steel riggers, who remove plates and bars to reach the interiors of the columns and chords (the chords are the top or bottom horizontal parts of a bridge truss). The bridge in its entirety is painted infrequently, but it's touched up daily. For the first twenty-seven years of its life, the bridge received only minor touch-ups. By 1965, however, corrosion sparked a program to remove the original paint and replace it with an inorganic zinc silicate primer and acrylic emulsion topcoat, an effort not completed until 1995. Chipping guns were originally used to remove rust, after which a primer coat and new orange paint were brushed on by hand—a slow, ineffective method. Chipping also cut into the steel, and the red lead paint did not penetrate completely. Rust quickly settled in. In the mid-1960s, sandblasting took over, followed by an undercoating of inorganic gray zinc primer left to cure for ninety days. A bonding agent of olive-green acid wash and, finally, two coats of International Orange vinyl are then applied. Automatic scaffolds, spray guns, and lead-free paint make the job safer and more efficient. Nonetheless, painting remains the single most costly maintenance item in the bridge budget.

Perhaps no feature of the bridge, other than its silhouette, is more noticeable than its color. And yet the choice was debated, challenged, rejected, and, finally, because of an artist's eye and persuasive skills, accepted.

CHAPTER 3

HIDDEN TOWERS

"When your car moves up the ramp, the two towers rise so high that it brings you happiness; their structure is so pure, so resolute, so regular that here, finally, steel architecture seems to laugh."

—*Le Corbusier, architect*

Though Le Corbusier's comment describes the George Washington Bridge, it also applies to that bridge's West Coast descendant, the Golden Gate Bridge, whose Art Deco towers dominate their mile-wide strait. Watching proudly over traffic, pedestrians, and ships below, they are likely the most recognizable bridge towers in the United States, if not the world. When they were built, they were taller than any building in San Francisco.

Yet the towers neither overwhelm nor daunt until one is beneath them. Their perfect proportions diminish their enormous scale; as they ascend, they recede into the sky, creating the impression from a distance of almost invisible support. To enhance their aesthetic appeal, fascia plates cover their bracing (Figure 37). The plates, which decrease in height from the ground up, are purely artistic but blend perfectly with the towers' stepped-back lines—Irving Morrow's idea. The scale of detail decreases with the towers' height, in contrast to the usual practice of enlarging details to increase emphasis, creating a soaring effect. The

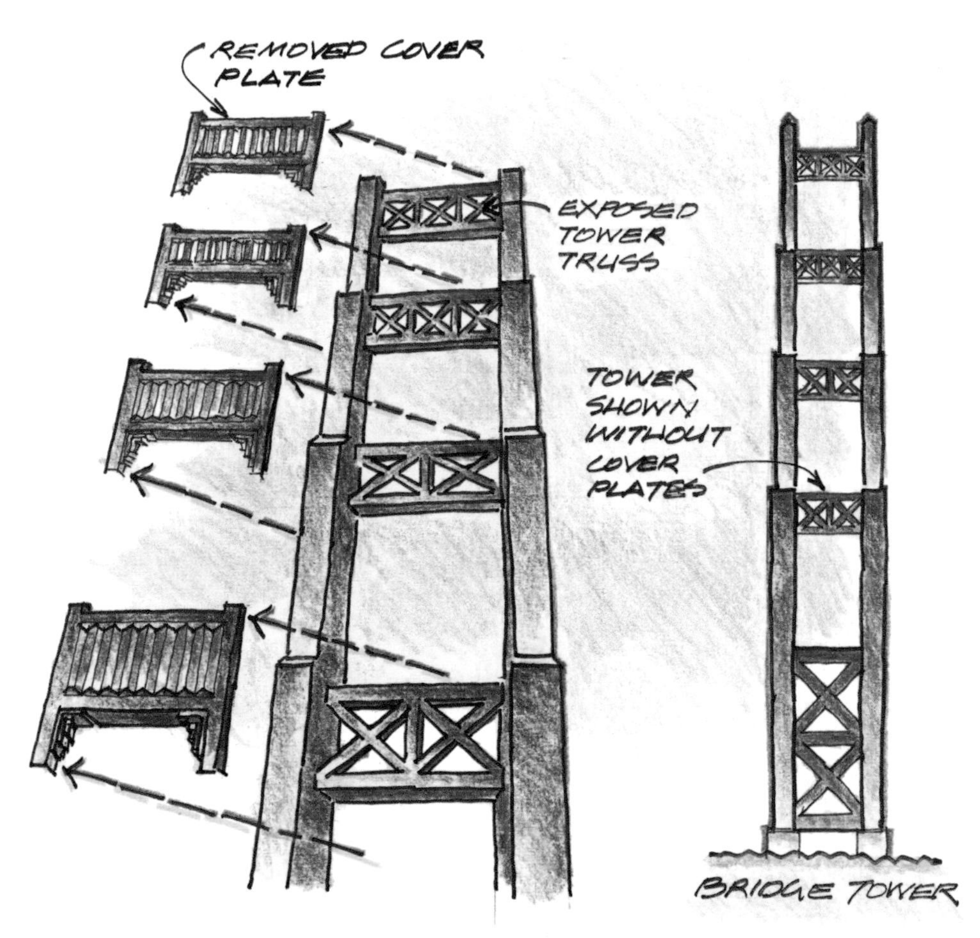

DRAWING SHOWING THE REMOVAL OF THE REDUNDANT STEEL PANELS ON THE TOWER'S HORIZONTAL TRUSSES

Opposite: Figure 37

towers' portals frame the sky, clouds, and fog. Fluted brackets echoing the vertical planar motif mark their upper corners (Figures 38–39). As noted in Chapter 1, the towers' indentations, or stepped-back feature, which heightens their clean, elongated effect, was the work of John Eberson. When Morrow reconsidered the towers in the 1930s, he made important artistic additions: vertical fluting, natural earth colors, tower portals of varying shapes, and the redesign of the horizontal portal bracing struts. The portals' frames brace the towers, rather than visually unappealing diagonal braces—the contribution of assistant engineer Clifford Paine.

Morrow summarized the designers' approach in 1930: the bridge's architecture, like its engineering, must belong to the present. The angular, clean, uncluttered look of the towers fuses function with aesthetics, strength with Art Deco principles. In a 1937 *Architect and Engineer* article, Morrow records that Strauss knew his bridge would be outstanding in two respects: in its magnitude as an engineering feat and in its domination of a beautiful and distinguished landscape. Hence Strauss's sensitivity to architectural design in the creation of what remains a remarkable and well-documented engineering achievement.

The elegant towers were quickly labeled the "Portals of the Future," and while they artistically dominate the bridge's form, they also have a crucial structural function. On top of each tower sit saddles—160 tons apiece!—that cradle the cables (Figures 40–41). Surprisingly, in warm weather, when the cables expand, they do not move in the saddles. Instead, the tops of the towers actually lean a few feet in each direction. To create transverse stability to balance the towers' longitudinal strength, they were cross-braced to act as a frame. In older bridges, such bracing took the form of an X. In an equally strong but artistically satisfying form, Morrow created a horizontal strut shaped like a truss but covered with ornamental steel fascia plates. The Golden Gate Bridge was the first bridge to eliminate a network of transverse bracing between its tower posts.

Figures 38 & 39

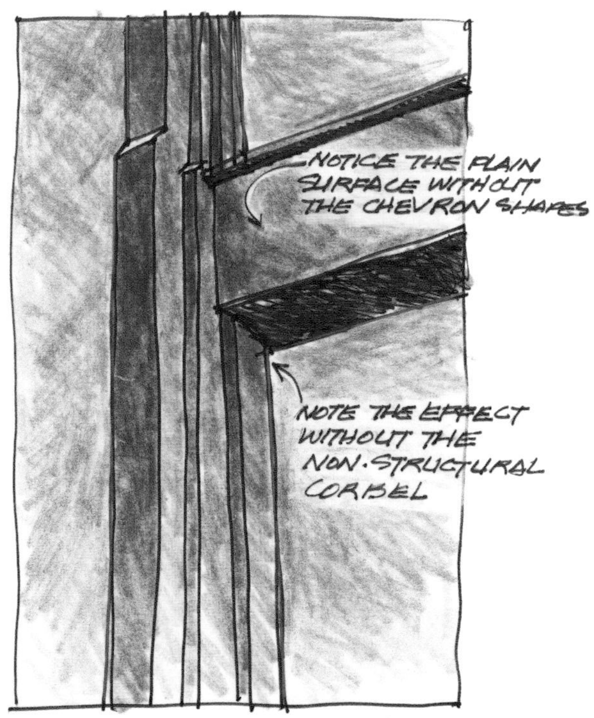

SKETCH TO EXPLAIN THE EFFECT OF SMALL SCALE ELEMENTS ON THE MAIN TOWER

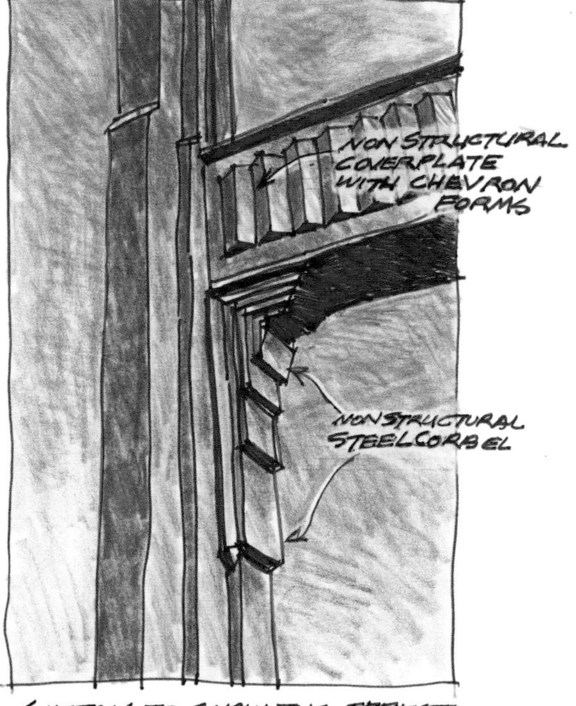

SKETCH TO SHOW THE EFFECT OF SMALL SCALE ELEMENTS ON THE MAIN TOWER

Figures 40 & 41

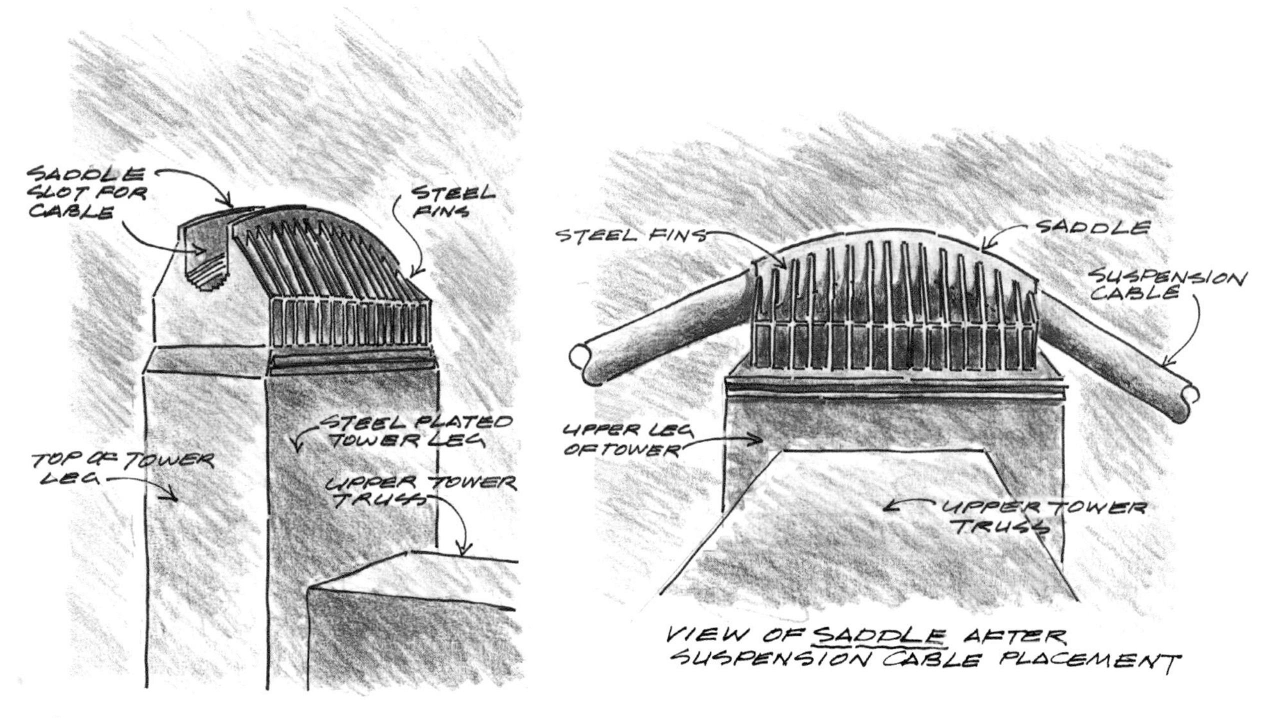

TYPICAL STEEL DECORATIVE PORTAL ELEMENT FOR THE UPPER CORNERS OF THE MAIN TOWERS
NOTE SCALE OF HUMANS NEXT TO THE PORTAL ELEMENT
RAILROAD FLAT CAR
ART DECO PORTAL ELEMENT PLACED ON A RAILROAD FLAT CAR FOR TRANSPORTING TO SAN FRANCISCO.

Opposite: Figure 42

Instead, its portal-braced towers look like a majestic doorway, elegant and airy despite their height. The steel towers, built for strength, have a design equal in grace to the cables.

On the towers' sides are long, slender flanges that grow narrower as the towers climb higher, forming a kind of Art Deco armor for the structure. Small-scale nonstructural elements also aestheticize the bridge: corbells used as insets to frame the horizontal steel girders connecting the towers, which also have nonstructural features, add to the balance of the design (Figure 42). The fascia coverplates, with their chevron forms, extend the Art Deco motif; chevrons even appear in the concrete of the tower foundations (Figures 43–45). But in their avoidance of applied decoration and unnecessary elements, the towers are the quintessence of Modernism. Simple vertical fluting, hand rails, lighting units, and other details all eschew ornamental elaboration. And as the towers rose, praise began. Sculptor Beniamino Bufano wrote Morrow in 1935 that the bridge "moves and molds itself into the general beauty and contours of the hills" due to the towers' clear engineering and architectural simplicity, as well as their color, "like red terra-cotta." The "structural simplicity" of the forms, he added, clearly defines their beauty. Others, too, praised the towers' harmonization of design, landscape, and light.

The towers' height alone establishes their majesty. At 746 feet, they dwarfed all other bridge towers built before them. The George Washington Bridge towers stand 595 feet, the Bay Bridge's a paltry 450 feet, the Brooklyn Bridge's only 273 feet (Figure 46). From the top of the Golden Gate towers to the roadway is nearly 500 feet; from the roadway to the water is more than 200 feet. Its towers also display an unusually close correlation between architectural and structural elements. Built as rigid frames made up of two shafts and cross-bracing without diagonals in the panels that compose the upper 500 feet of their height, each tower contains 22,000 tons of steel, about 10 percent more than was

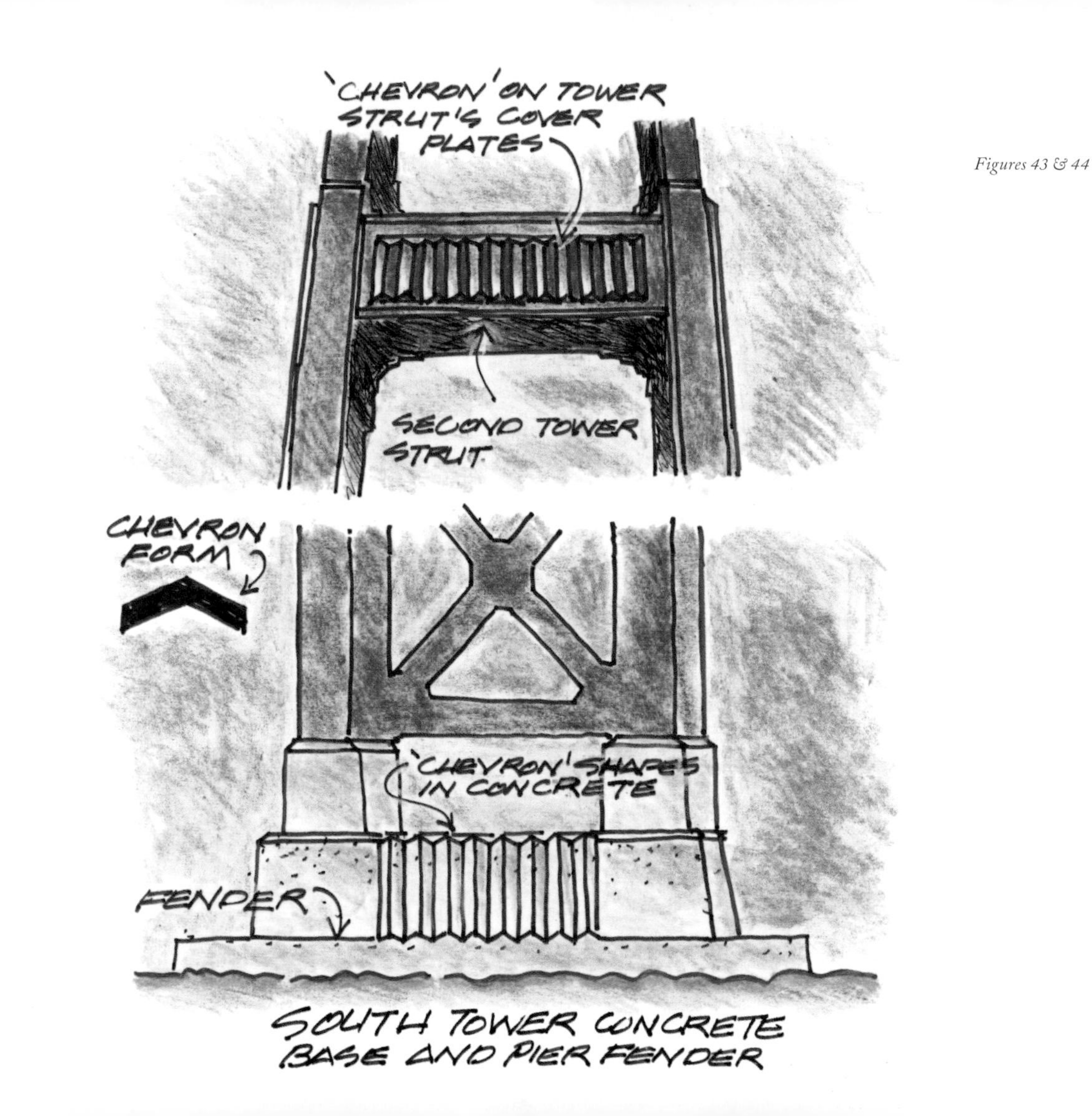

Figures 43 & 44

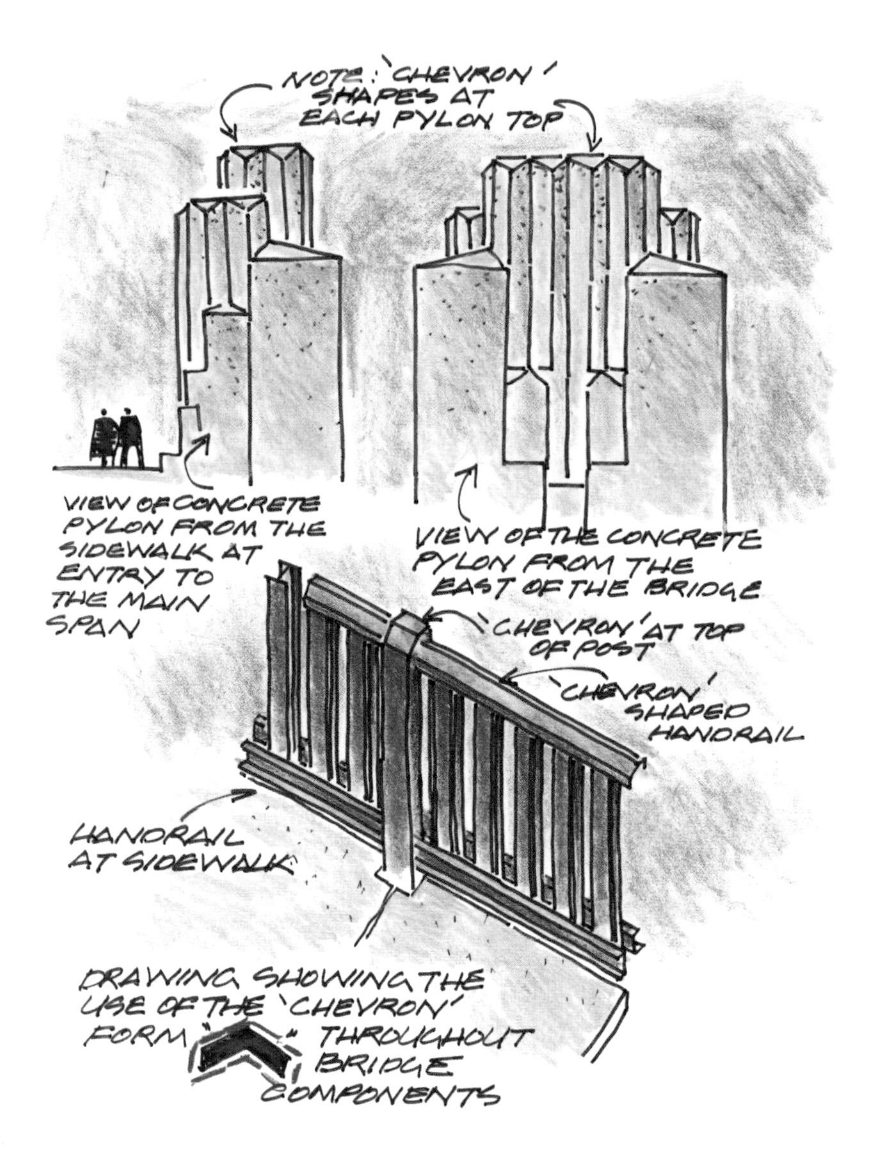
NOTE: 'CHEVRON' SHAPES AT EACH PYLON TOP
VIEW OF CONCRETE PYLON FROM THE SIDEWALK AT ENTRY TO THE MAIN SPAN
VIEW OF THE CONCRETE PYLON FROM THE EAST OF THE BRIDGE
'CHEVRON' AT TOP OF POST
'CHEVRON' SHAPED HANDRAIL
HANDRAIL AT SIDEWALK
DRAWING SHOWING THE USE OF THE 'CHEVRON' FORM " " THROUGHOUT BRIDGE COMPONENTS

Figure 45

HEIGHT OF
TOWER 746'
HEIGHT OF
TOWER 595'
HEIGHT OF
TOWER 273
BROOKLYN
BRIDGE
GEORGE WASHINGTON
BRIDGE
GOLDEN
GATE BRIDGE

Opposite: Figure 46

used in each tower of the George Washington Bridge. The play of light, too, distinguishes the towers of the Golden Gate Bridge. To mirror the interplay of light and dark on the Marin hills, Morrow captured sunlight via geometric planes on the bridge. The resulting shadows add volume and dimension.

In November 1936, when the final piece of the single center span was put into place and the two sides of the Gate connected for the first time, with suspender ropes connected and taut, the bridge looked like "a great red harp, hung in the western sky" according to the bridge historian John van der Zee. The span is in proportion to the towers (the ratio is 1 to 3: the roadway is one-third the towers' height, following the Renaissance principle of thirds), so the bridge's actual size and the distance it covers are not apparent. The center lines of the tower shafts are 90 feet apart—the desired distance of the cables across the roadway—so a distinct but parallel separation is achieved. The forms harmonize, the almost airborne height of the road creating drama for ships passing beneath and marking every arrival and departure with significance.

The towers' stepped motif, as noted earlier, had not been used on a bridge before, but had been used on New York's Chrysler Building and proposed Empire State Building (Figure 47). Traditionally, bridges were bulky in part because their towers were larger at the top than at the base to preserve uniformity as the eye adjusted for distance and perspective. Where great heights are involved, scale of details usually enlarges as distance from the ground increases, emphasizing detail at the expense of mass. The opposite approach, however, was employed on the Golden Gate Bridge: the towers' scale of details decreases with height to emphasize their vertical silhouettes and grace. This partly came about because advances in metallurgy allowed the use of a steel cell tower structure (essentially steel cubes placed on top of each other) rather than solid steel beams, creating a lighter, taller, more slender structure. Additionally, the strut spacing of the

Below: Figure 47

towers, diminishing progressively from roadway to tower top, accentuates their height, a radical departure from traditional design. The design of the pylons, symmetrically grouped in pairs at the two anchorages, echoes the stepped-back motif. The close correlation of architectural and structural requirements resulted in a remarkable design that balances structural needs with proportional form.

A hidden feature provides the towers' support, and proved to be the most difficult element to build: the pier foundations. The Marin County pier, close to shore, presented few problems. The San Francisco pier, built 1,100 feet into the open sea, unprotected from the elements and sitting in 60 feet of water, was another story. Initially, a trestle bridge was built out to the spot where the fender (a concrete ring built to enclose the site for the construction of the tower base) was to be constructed underwater. Not long after it was built, a ship, thrown off course in the fog, crashed into the pier and nearly destroyed it. Then gale-force winds carried off nearly 800 feet of it. It was rebuilt. But when the south pier's caisson (a giant underwater steel chamber where men can work protected from water pressure by compressed air) was towed into position in a large concrete fender, a heavy sea swell caused it to bob like a cork.

EMPIRE STATE BUILDING ON FIFTH AVE · NEW YORK · CONSTRUCTED IN 1931.

Though the fender was meant to protect the caisson, the caisson now threatened to destroy the fender. The caisson was towed out of the fender, and the fender itself was enclosed as a cofferdam (an open-air structure that keeps out the sea to allow underwater excavation and foundation construction). Of course, no one recalls these structures today, yet they were instrumental in forming the foundations of the bridge.

A final hidden element is the small internal elevators in each tower, extending from deck to tower top, which enable workers to reach the tower saddles for repairs. Originally, they were also intended to ferry riders to glass-domed observatories at the top, but budgets forbade such extras.

The towers are the most prominent and distinctive feature of the bridge, vertical monuments to beauty and design. Their construction was the result of technical advances in the use of steel, as well as the fruit of the imagination of their engineers and architects.

CHAPTER 4

MONUMENTS IN THE AIR: THE PLAZA AND TOLL BOOTHS

"The magnitude of the project calls for architectural treatment elevated in spirit and scale."

—*Irving F. Morrow, bridge architect*

Bridge approaches are entranceways, and they should signal both the nature and the form of the space to which they grant access. For a structure as monumental as the Golden Gate Bridge, they had to be grand, impressive, and as unforgettable as the bridge itself.

The 1930 bridge design originally called for plazas at both the north and the south approaches. The San Francisco plaza was to have exhibits and memorial halls, as well as space for bridge personnel and bathrooms. The Marin portal was to be smaller and less elaborate but still magnificent, as outlined in Irving Morrow's drawings.

Monumentality mixed with beauty was Morrow's challenge for the south plaza, and visual drama dominated his design. Bernini's colonnade at St. Peter's Basilica in Rome (Figure 48) provided inspiration: orderly space narrows only to open dramatically to the bridge itself. The entry portals were to be faced with terra-cotta and decorated with murals and sculptures. Paving and planting inside the plaza were part

BIRD'S-EYE VIEW OF SAINT PETER'S CATHEDRAL COLONNADE · ROME.

Opposite: Figure 48

of the aesthetic scheme. Broad-angled walls narrowing toward the bridge entrance would be painted in polychrome, increasing in richness as one drew closer to the deck. The pylons, the last solid land structures seen by drivers and pedestrians, would be painted in tones culminating in pure metallic gold as they approached the towers, reflecting, according to Morrow, "progressively enriched color up to the final monumental masses in pure metallic gold—veritably as well as symbolically the Golden Gate." The south plaza's west side was to have an extended, high building to act as a windbreak to the prevailing strong westerlies, leaving the low east side open as a loggia with views across park spaces to the Bay's upper reaches (Figures 49–50). Under the "San Francisco Windbreak," as Morrow termed it, were to be a museum and a display of California products.

Strauss's original 1924 plan included an entryway modeled on the Arc de Triomphe to match the Parisian motif of the bridge itself: a sort of Champs-Élysées built across the water, with towers evoking the Eiffel Tower. John Eberson outlined a second entryway plan in the San Francisco *Chronicle* of 5 October 1930: a horseshoe-shaped structure with columns appeared at the San Francisco side, with the road cut through its middle. Realignment of the roadway by federal authorities, however, necessitated redesign of the San Francisco plaza, adding cost—the reason most of these elaborate plans were halted.

Money was, in fact, a stumbling block from the beginning. In 1930, a bond issue to begin construction was set before San Francisco and Marin voters, but many opposed the idea. The Citizens' Committee Against the Golden Gate Bridge ran radio and newspaper advertisements, and canvassers went door to door to oppose the project. They argued the bridge would be impossible to build, would destroy the site's beauty, would alter Sausalito's wonderful isolation, would be an earthquake hazard, would bottle up the harbor if destroyed by an enemy, and could not be supported by tolls paid by Marin County's sparse population.

Figure 49

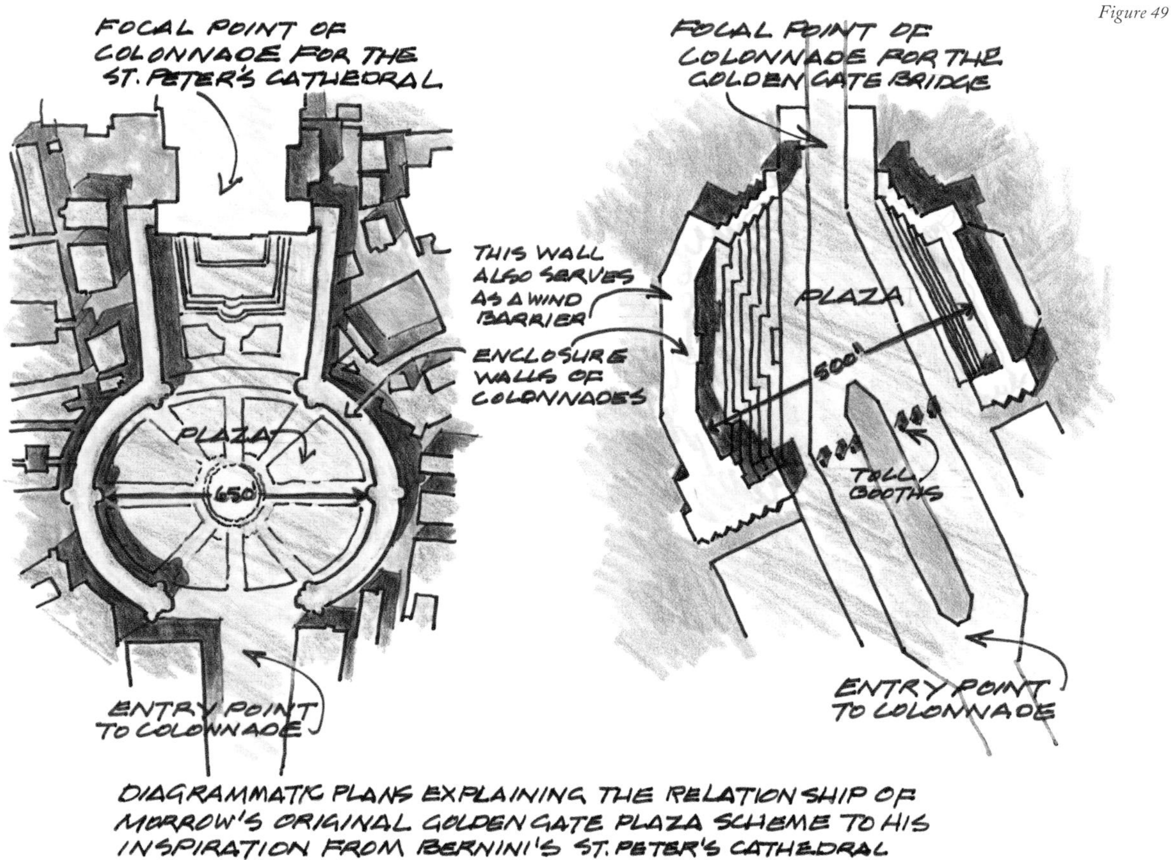

DIAGRAMMATIC PLANS EXPLAINING THE RELATIONSHIP OF MORROW'S ORIGINAL GOLDEN GATE PLAZA SCHEME TO HIS INSPIRATION FROM BERNINI'S ST. PETER'S CATHEDRAL COLONNADE IN ROME.

Figure 50

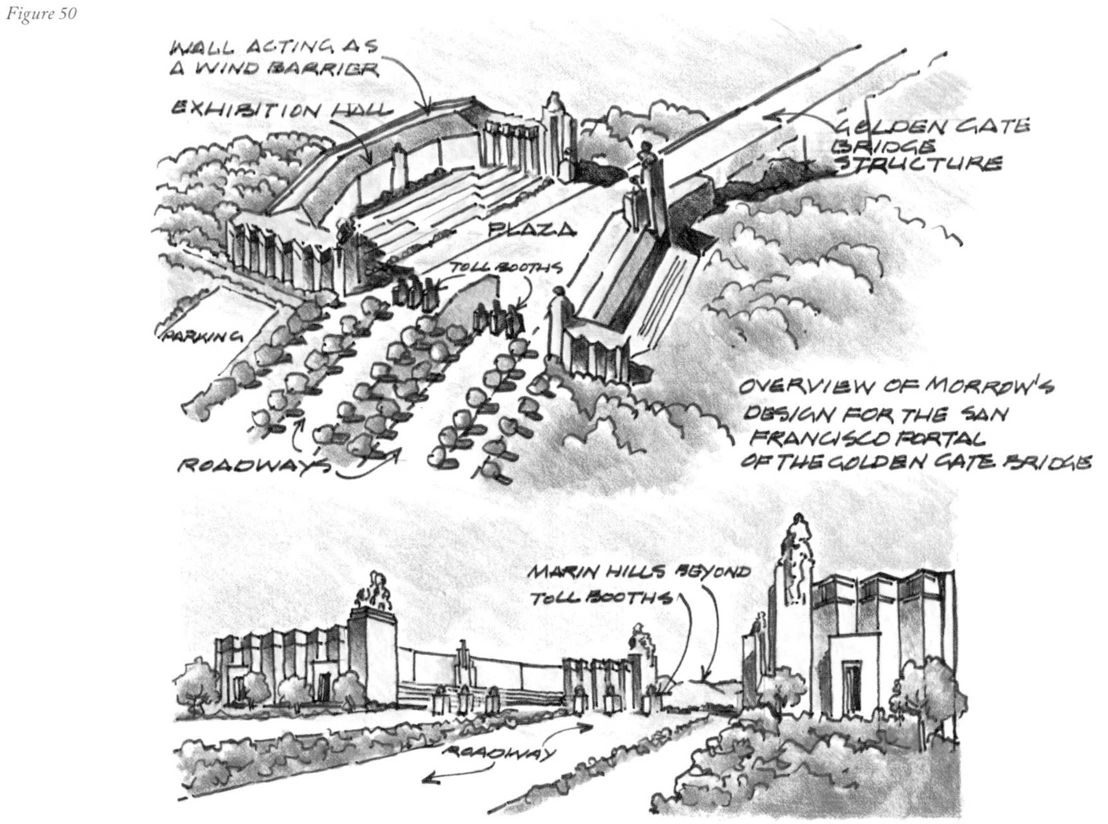

SAN FRANCISCO PORTAL FROM APPROACH ROADWAY

Furthermore, it was the Depression, and public money was limited.

The "yes" side, however, carried the day because more understood the bridge's economic and transportation benefits; on 30 November 1930, the bond issue passed. Yet it was immediately caught up in lawsuits launched by Southern Pacific–Golden Gate Ferries and, later, the railroads. Meanwhile the Depression worsened and the bond market remained dormant. In desperation, Strauss and a delegation went to see Bank of America founder A. P. Giannini. He and the bank agreed to buy the bonds; with financing in place, the first shovel of dirt was turned on 5 January 1933.

Financial constraints, however, continued to hamper the work, especially on the south plaza, and by 1936 many alterations had to be made to Morrow's original conceptions. Even his worry about wind met with little sympathy, as did his wish to use color to offset the constant monochromatic gray fog and cloud. Only in this way, he argued, could the structure be constantly alive, its irregular but organized forms sheltering it from the winds yet remaining open to the city and converging on the bridge proper. His plaza buildings, lit from above and the outside, would uniquely display the proposed murals and sculpture. But his uplifting ideas were not to be: on 11 March 1937, a distressed Morrow wrote to Strauss that "the Plaza is assuredly going to appear bare." On 14 June 1937, Morrow concluded that the toll plaza had "suffered so many cuts and arbitrary changes that most of it can not now be taken very seriously." The plaza became utilitarian rather than monumental and never held exhibition halls, a museum, or sculpture.

Like the plazas, the toll booths commanded attention from the first because of their elaborate structure. Their earliest conception formed part of Strauss's Arc de Triomphe scheme, grand and historical. Eberson had his own idea: a majestic entryway for the booths and a majestic monument dividing the twelve toll lanes. Morrow had still another view: a modernistic entranceway. His 1930 drawings depict an extended portal building incorporating toll booths on the San Francisco side and a smaller one on the north. The 5 October 1930 *Chronicle* depicted Eberson's plan—a horseshoe structure with columns on the San Francisco approachway, emphasizing grandeur. A 1932 scheme centered the elongated, Art Deco–themed toll booths in the middle of the plaza, eliminating any divider. Money, however, limited the final overall design, in which narrow, streamlined booths combined sand-colored steel panels with rounded glass windows.

By 1937, the south plaza's originally planned walls, rails, and pylons, designed to unite the entire composition and set off the buildings, were omitted except for a utility fence, creating problems for both traffic handling and administrative functions. Smaller office space would impair efficiency and maintenance. What was built by opening day was utilitarian, unaesthetic, and lacked symmetry. It was also the least costly option. The plaza became pedestrian, and the west side housed administrative offices, garages, storerooms, shops, and a powerhouse. Fourteen, not twelve, toll booths stretched across the plaza, designed in a powerful "airflow" style with curvilinear corners and an Art Deco character.

In the bridge's early years, the booths had key boxes with a key for each classification of traffic. When the toll collector pressed a key, the toll was displayed over the driver's lane. The toll in 1937 was fifty cents (Figure 51), the price of a decent meal. In those days, monies were bagged and deposited in a special chute that led from each booth directly into a money room below the roadway. As bridge maintenance costs increased over the years, tolls rose, but they remain the bridge's main source of revenue. Now, at $5 a toll, almost $250,000 a day is collected; the FasTrak system, with computerized cards, was installed in 2000.

Incidents both comic and dangerous occur at the fourteen booths. Collectors have been handed everything from dead fish to live kittens, pizzas, fruit, and

Below: Figure 51

even loaded guns. Drivers have proffered money in their teeth, behind their ears, and between their toes. Collectors have witnessed prison escapees, battling lovers, and robberies. Drivers have fallen asleep at the booth, while others backed up traffic to fix their makeup. Frequent drivers argue that they should not pay tolls since the bridge was paid for long ago. A woman who had just shot her boyfriend pulled up to a booth with him bleeding in the car and handed the gun to the collector, who, thinking it was a joke, accidentally fired a shot into the floor of the booth—an incident surpassed only by the drunk who passed out at a booth in his locked car, which no one could open, or the go-go dancer whose friendship

ART DECO LETTERING FOR THE TOLL SIGNS AT THE BOOTHS.

with a collector led to a quick rendezvous in his booth early one morning. Naked drivers also appear, although generally not at peak commuting hours. Those who cross the bridge without money must sign a promissory note in an adjoining office and leave collateral, which has included radios, dentures, a toupee, and an artificial arm. All this should not be surprising given that more than seven hundred cars an hour often pass through a single toll lane. A good collector, officially called a "bridge officer," must be a teller, traffic policeman, and public relations agent.

Yet the toll booth plaza remains one of the most unsettled aspects of the entire bridge project. The bridge design seemingly fixed the character of the booths, but they have been constantly revised. In May 1980, bridge engineers felt it was time to upgrade and redesign the booths, and they came up with a squat, sentry-box replacement. Public outcry followed. Bridge directors were reminded that the original plaza was intended to be the portico of a classical temple, a grand entrance to San Francisco. After a new competition, MacDonald Architects' 1980 redesign echoed the curved glass and Art Deco structure of the original booths (Figure 52), using their elliptical shape but including flat glass protected by rounded horizontal steel bars for stability and security (Figure 53).

Work on the plaza continues. In 1972, Congress authorized the Golden Gate National Recreational Area, protecting the parkland and historic sites surrounding the bridge. And in 1980, a federal study concluded that the area should be nominated a National Historic Landmark district. But not until 1989, when the U.S. Army announced that it would vacate its Presidio post by 1995, did action to preserve the structures surrounding the bridge begin. In 1994, a redesigned plaza was proposed. It would increase visitor access via multilingual interpretative facilities and exhibits featuring the bridge, its importance in coastal defense, and area natural history. Some maintenance and administrative facilities would relocate, freeing land and space for public access and scenic vistas. Some of this work has been completed over the years, but much still needs to be done.

The area below the bridge is as important as that above. Among the most interesting preservation targets is Fort Point (Figure 54). Here, in 1794, the Spanish built the Castillo de San Juan, adjacent to the Presidio they established next to the Golden Gate crossing. The batteries and fire-control stations that ring the point date from the Civil War to World War II. Fortifications within the post once housed the two earliest Endicott rifle batteries in the country. Built in 1861, the fort itself is actually a red-brick Civil War–era army blockhouse

Figures 52 & 53

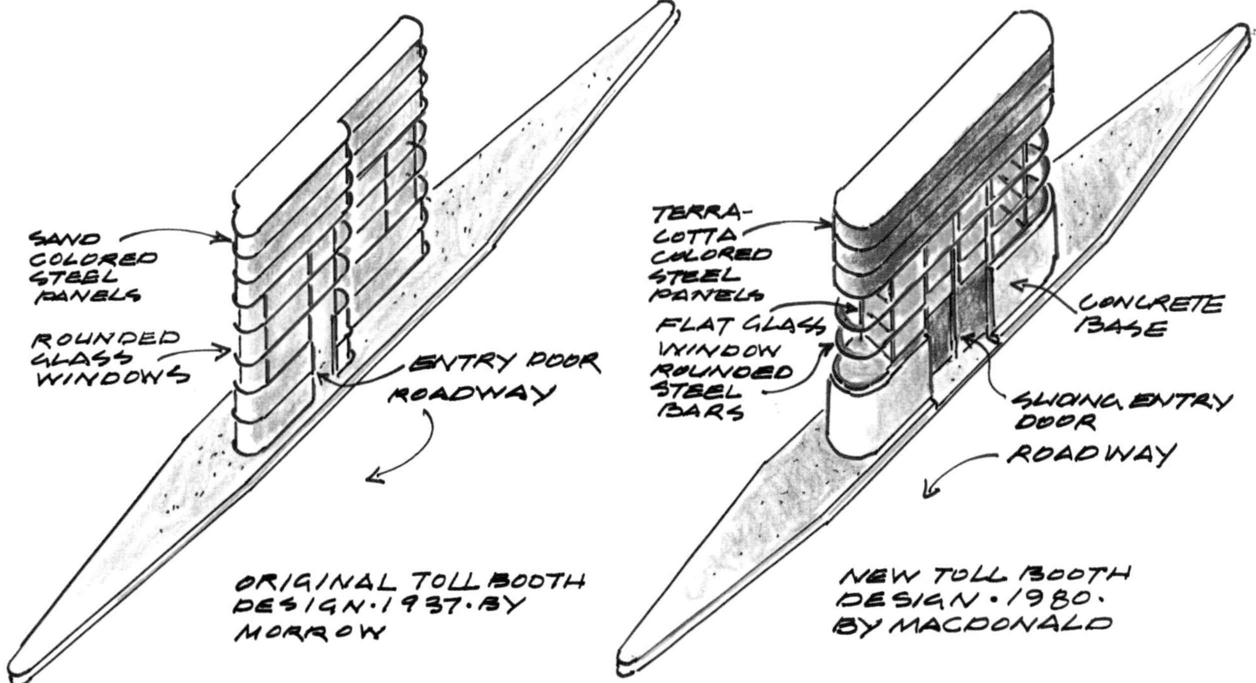

Figure 54

FORT POINT

Figure 55

SKETCH SHOWING IRVING MORROW'S STUDY
AT FORT POINT WITHOUT THE EXISTING ARCH

Figure 56

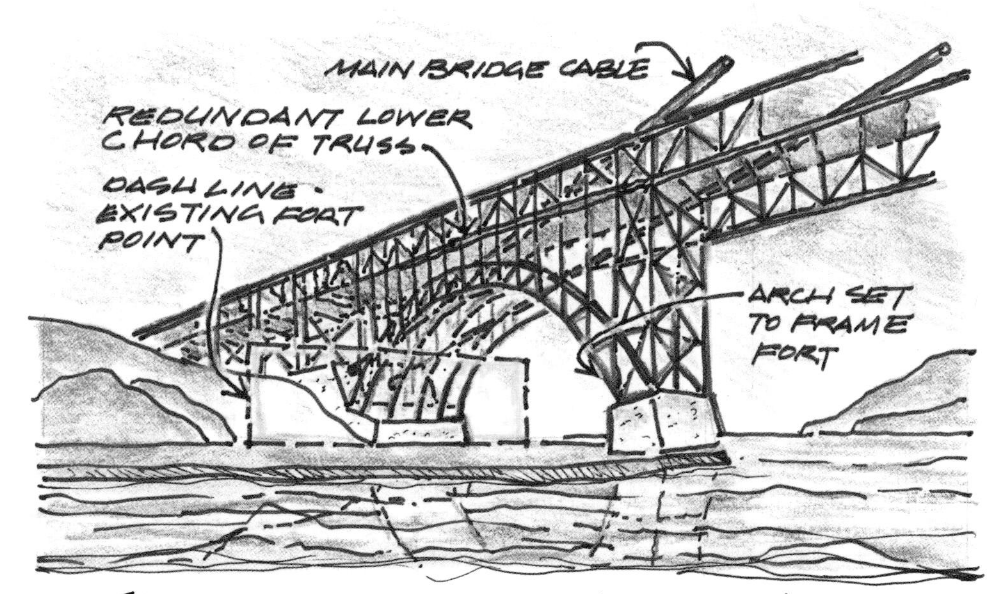

SKETCH SHOWING IRVING MORROW'S STUDY FOR THE EXISTING ARCH OVER FORT POINT

NOTE: PYLON NOT DEVELOPED AT THIS STAGE OF THE SOUTH VIADUCT DESIGN

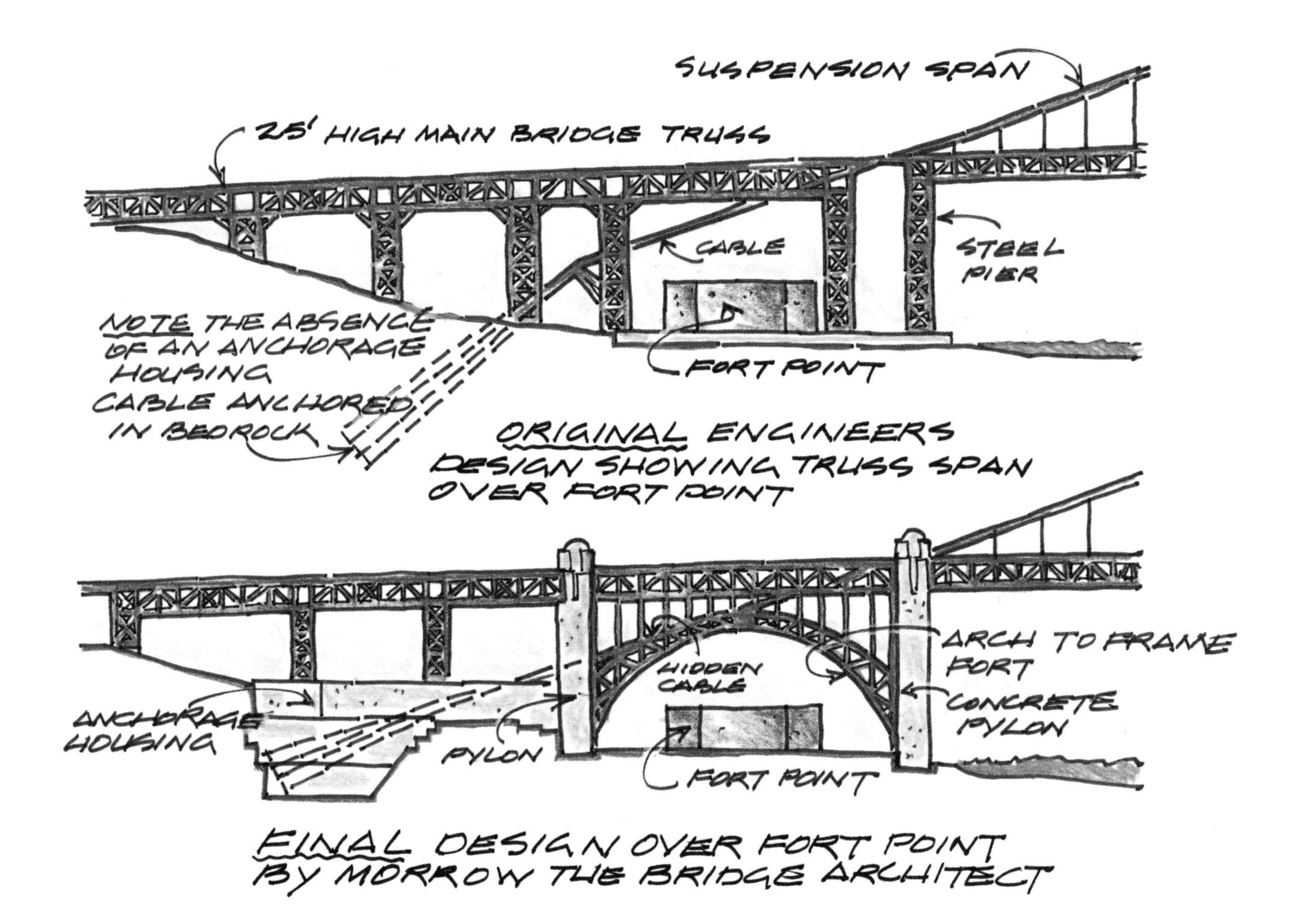
SUSPENSION SPAN
25' HIGH MAIN BRIDGE TRUSS
CABLE
STEEL PIER
NOTE THE ABSENCE OF AN ANCHORAGE HOUSING
CABLE ANCHORED IN BEDROCK
FORT POINT
ORIGINAL ENGINEERS DESIGN SHOWING TRUSS SPAN OVER FORT POINT
ARCH TO FRAME FORT
HIDDEN CABLE
CONCRETE PYLON
ANCHORAGE HOUSING
PYLON
FORT POINT
FINAL DESIGN OVER FORT POINT BY MORROW THE BRIDGE ARCHITECT

Opposite: Figure 57

patterned on Fort Sumter in South Carolina. The fort never saw action; demonstrations of its cannons' power showed that they could never fire far enough to hit a barge in the channel. Designated a National Historic Site in 1970 and the Bay Area's first national park site in 1971, Fort Point, both literally and metaphorically, is a meeting place of biological systems, explorers, settlers, and soldiers.

The original plan for the bridge's south cable anchorage meant the destruction of Fort Point. Strauss, however, recognized its historical importance and located the anchorage forms behind the building, a costly aesthetic "detour" that necessitated construction of a 319-foot steel arch with flanking concrete pylons to support the bridge floor, hide the main span's cables, and protect the fort (Figures 55–57). Today the fort is handsomely set off by the towering archway, which balances the north anchorage housing and creates a symmetrical design.

Attention to the plaza and toll booths, plus protection of Fort Point, demonstrates the bridge engineers' and architects' awareness of the impact of the bridge on its adjacent areas and the need for artistic consistency. The integration of these elements echoes the overall unity maintained by the approaches and the principal structure.

CHAPTER 5

LIGHTS, WIND, ACTION!

"The object is to reveal aspects of a great monument that are unsuspected under the conditions of natural, or day, lighting."

—Irving Morrow on lighting the Golden Gate Bridge, 1936

How does one light a bridge that is 1½ miles long? That was what chief engineer Joseph Strauss asked his architect, Irving Morrow. His question was itself unusual, since electrical engineers usually dealt with lighting questions. But Morrow's answer was equally unusual. Seeking something artistic as well as practical, Morrow submitted a detailed plan in the spring of 1936. He thought lights should be placed for dramatic effect, avoiding a uniform distribution that would seem artificial and "compromise the effect of size"—this last remark indicating an architect's rather than an engineer's sensibility. Morrow felt similarly about illuminating the towers: if lighting diminished in intensity as the towers rose (Figure 58), the gradation would give them a "sense of soaring beyond the heights susceptible to facile illumination." In short, Morrow wanted to combine relatively low intensity with constant changes of light. He approached lighting as paint.

Light and shadow were used to emphasize the bridge's size and scale. At night, light paralleled Morrow's

NIGHT LIGHTING ON THE TOWER • TYPICAL EACH TOWER

Opposite: Figure 58

use of vertical fluting to accent the towers' height and grace. He wanted the tower bases to be enveloped in a mellow glow, with the tone gradually disappearing into the night. The cables were suggested by diffusion from roadway lighting and from the towers. Approach viaducts vanished into the darkness. The roadway, however, needed a strong, continuous band of light, particularly to combat fog, but he still wanted it to be lit decoratively.

The new sodium-vapor lamps chosen by Morrow, while expensive, cut through fog and bathed the bridge in a transforming orange glow, unifying all its elements and avoiding a spectacular effect that could have compromised the bridge's dignity. The lights' placement, especially the diffused light inside the tower portals, creates a powerfully moving, if spectral, sense of illumination. Morrow's goal—one sweep of diffused light from sea to towers—was achieved. His vision of the lighting was, along with the bridge's color, perhaps his most dramatic contribution.

But what should the lights themselves look like, and how would they illuminate the roadway as well as the superstructure? Design was crucial and ever-changing. Preliminary 1929 plans had bordered the bridge's pedestrian walkways with an elaborate metal grillwork railing and street lamps that suggested turn-of-the-century elevated train stations. Morrow, who had hiked in the nearby Marin foothills and understood that the bridge would be the object and vantage point of spectacular views, eliminated this "embroidery" to allow drivers and pedestrians full access to vistas on either side. He simplified the railing into a line of posts spaced slightly farther apart than usual, a barrier that dissolves into a scrim and does not impede views. Replacing the street lamps were slender angular standards that curved over the roadway and echoed the towers' angularity, contributing to the bridge's uncluttered, modern look while maintaining its Art Deco character (Figure 59).

The government mandated additional lights: navigation lights, revolving air beacons atop each tower,

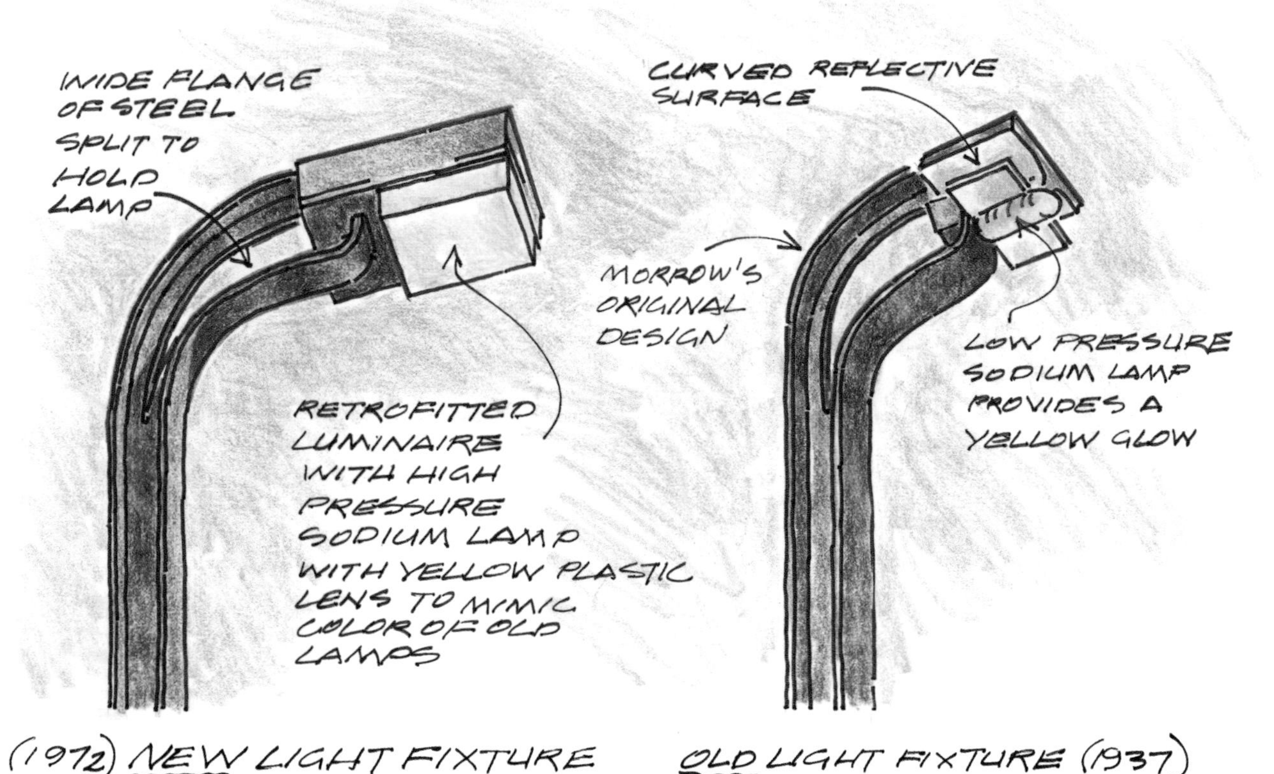

(1972) NEW LIGHT FIXTURE OLD LIGHT FIXTURE (1937)

COMPARISON OF OLD AND NEW FIXTURES

Opposite: Figure 59

lamps on both bridge piers, constantly illuminated midchannel lights, and obstruction lights on each cable, plus foghorns, two at the center span and two at the San Francisco pier. An emergency power source was therefore needed, and an 11,000-volt electrical substation was built for the toll plaza. Located high above the water at the center of the main span, the bridge's foghorns are the only ones in the world toward which boats may steer instead of avoiding, because they are set hundreds of feet above the channel.

Today, there are 128 high-pressure sodium lights from abutment to abutment, and 24 low-pressure 35-watt sidewalk lights around the towers. Above the roadway, 12 decorative 400-watt high-pressure sodium lights surround the towers, and 12 more sit at the roadway below the towers. Two 700-watt air beacons sit atop each tower, and 8 lights illuminate the bridge's midspan.

"A cable-supported bridge is subject to wind-induced drag (the static component), flutter (the instability that occurred at Tacoma Narrows), and buffeting (where gusts 'shake' the bridge)."

—*Mark Ketchum, bridge engineer*

Wind is perhaps the greatest constant danger to the bridge. Yet discoveries about the effects of wind on suspension bridges made its current design possible. Engineer Leon Moisseiff's wind-stress distribution theory, calculated by balancing sideways movements or horizontal deflections in the cable with those in the bridge's stiffening truss (the beam framework supporting the roadbed), allowed him to project load factors and figure out which portion of the total wind load was carried by the truss and which by the cable. Moisseiff showed that as much as half of the wind pressure could be absorbed by the main cables in a long suspension bridge and transmitted to its towers and abutments. The deflection of the truss and the cable would offset

Opposite: Figure 60

each other, keeping the bridge in equilibrium. A properly balanced bridge with its suspenders adjusted would not, therefore, be damaged were it flexible enough to bend and sway in the wind, and even bridges in high-wind areas like the Golden Gate need not be heavy or rigid. Long, light, narrow spans could survive, and would also be cheaper and quicker to build.

But how would the Golden Gate Bridge, specifically, handle wind? Charles Ellis, the original structural engineer, considered how its tower legs and diagonals could carry wind shear, but it was found that the towers' stiff legs contributed to, rather than solved, the problem. And wind caused problems in construction. After the north tower had stood finished for more than a year, wind and fog damage cost an estimated $24,000 to repair, mostly in repainting costs. Additionally, salt air whipped up to high speed corroded and undermined the steel surfaces. The Bay Bridge, 5 miles inside the Bay, did not suffer similar effects.

On 9 February 1938, ten months after the bridge opened, a gale-force storm hit the Golden Gate with gusts of 78 miles per hour that sometimes struck broadside. The center of the roadway toward the Marin tower was deflected 8 to 10 feet from its normal position and the suspended structure undulated vertically, reported the bridge's engineer, who had ventured out. Yet the bridge held. On 11 February 1941, a 60-mile-per-hour wind caused the towers and roadway to bend from their axes by nearly 5 feet. And on 1 December 1951, another storm forced the closure of the bridge for the first time: 69-mile-per-hour winds rippled the roadway, with one side pitching 11 feet higher than the other. Steps to reinforce the bridge floor followed, though there was little damage: some suspender ropes had pulled their socket plates out of shape and some light standards were bent back virtually to the roadway. Upgrading began with nearly 5,000 tons of lateral bracing steel. The bridge was closed again in the winters

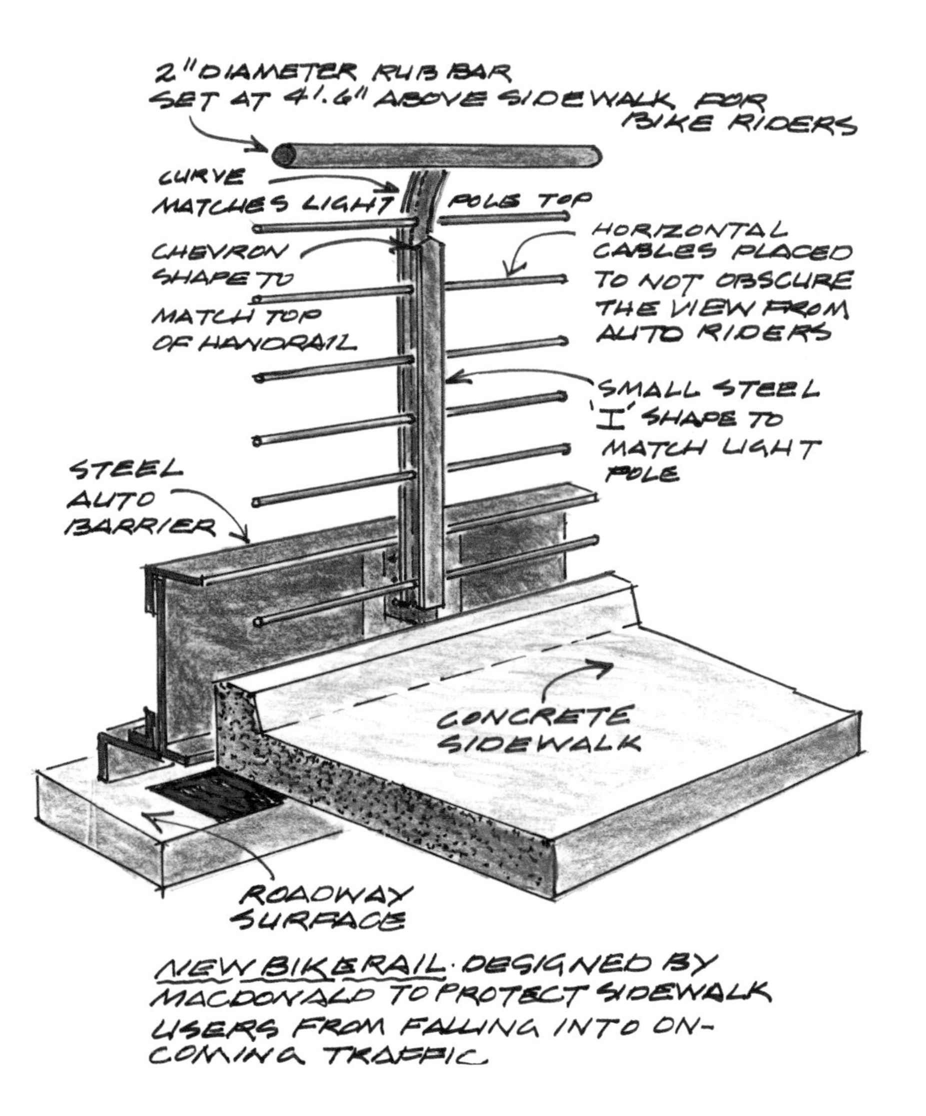

NEW BIKE RAIL · DESIGNED BY MACDONALD TO PROTECT SIDEWALK USERS FROM FALLING INTO ON-COMING TRAFFIC

Opposite: Figures 61 & 62

of 1982 and 1983, when continuous storms threatened the Bay Area. The dangers of high winds, treacherous height, and slippery fog condensation to workers were clear to Strauss and his associates. During construction in the summer of 1936, they placed a trapeze-type safety net under the skeletal roadway steel, perhaps the most dramatic safety measure in the history of civil engineering. It stretched 10 feet wider than the roadway on either side and bellied down 60 feet. Strauss realized, too, that increased security would encourage the men to work faster and more confidently. And the net saved nineteen lives—in contrast to the Bay Bridge, where twenty-two men died during construction.

Although there were injuries ranging from broken bones to burns and bruises, there was not one serious accident or fatality—until October 1936. That month, a bridgeman missed his footing, fell into the net, and broke his back when it bellied and fell against the rocks on the Marin headlands. Four days later, on 21 October 1936, the top of a stiffleg derrick pulled loose near midspan, causing the entire crane to come apart; a support beam slammed into a worker and crushed him to death. Several months later, an accident on the steel arch over Fort Point sent three men to the hospital. And in February 1937, during roadway paving, a platform below the roadway collapsed and sent men and debris into the net, which broke free and fell 200 feet into the water. Ten died.

Yet from the start, bridgeworker health was a concern. Special diets to counteract dizziness were prescribed for high-steel workers. Tinted goggles were issued to prevent "snowblindness" from sun reflection, and men had to wear hard hats and use safety lines on pain of dismissal. Hearing damage was an issue, and the noise from rivet guns, derrick engines, compressors, and steel slapping against steel made work and communication difficult, if not impossible.

Health and the bridge, with its crowds of fitness-seeking joggers and cyclists, have always been linked. But the latter long lacked protection; although there was never a fatality, cyclists often tripped, fell, or collapsed onto the roadway because no proper railing protected them. The author of this book designed a railing of horizontal steel cables (Figure 60) that establishes a "shy distance" (the space between a biker and the roadway) of 18 inches. Anything less would be too dangerous; anything vertical would block the view. The railing, installed in 2004, is largely invisible to drivers; it also continues the Art Deco motif, with a chevron form topping a vertical bar set on posts that hold the horizontal cables in place. This matches both the top of the handrail and the chevron design repeated throughout the bridge.

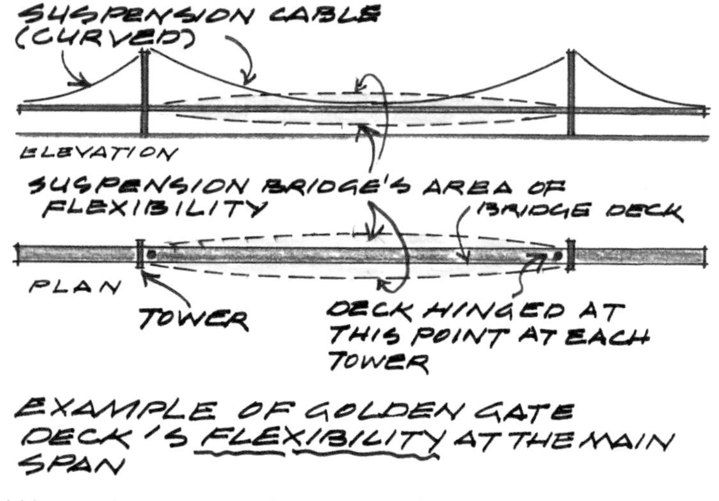

After wind, earthquakes are the greatest hazard to the bridge. How does the Golden Gate Bridge survive them? Its form, an elastic suspension bridge, certainly helps. Its ability to sway (Figures 61–62) permits it to respond to movements both laterally and longitudinally, and its lace girder chords and open trusses deflect not only wind but the ground energy and shifts of an earthquake. The solidity of its foundations allows it to withstand the tremendous sea swells occasionally caused by earthquakes. By contrast, the fixed form of the Bay Bridge does not protect it: in the 1989 Loma Prieta earthquake, registering 7.1 on the Richter scale and centered

ILLUSTRATION SHOWING MAIN SPAN IN A <u>STABLE</u> POSITION

ILLUSTRATION SHOWING MAIN SPAN <u>OSCILLATING</u> · FROM THE 44 MPH WIND · JUST BEFORE TOTAL FAILURE

TACOMA NARROWS BRIDGE · PUGET SOUND · WASHINGTON
BRIDGE FAILURE OCCURRED IN 1940.

Opposite: Figure 63

60 miles south of San Francisco, one of its spans collapsed. Other area bridges were also damaged, but not the Golden Gate Bridge. Why not? One reason might be its ratio of width to length, which protects it from collapses such as the 1940 Tacoma Narrows Bridge disaster (Figure 63).

Still, early-1990s seismic reports pointed out that in a severe earthquake, its north and south approach viaducts would collapse, its suspension span likely would ram its towers, and the Fort Point arch would uplift and move. A five-year, $175 million seismic retrofit (with the author as the lead architect) to reinforce concrete piers, steel tower shafts, and main cables is now well underway, focusing on structures that move with, rather than resist, seismic forces.

CHAPTER 6

THE TWENTY-FIRST-CENTURY BRIDGE

"Because of its excellent design, its history of significant structural improvements, and ongoing maintenance program, the Bridge has a life span estimated at 200 years."

—*Seismic Retrofit Report, 1994*

SEISMIC UPGRADES

Retrofits at the Golden Gate Bridge began more than fifty years ago and have included three major modifications. The first began after the howling gales of 1 December 1951, which caused minor but noticeable damage. Clifford Paine, one of the bridge's original engineers and head of a study on the 1940 Tacoma Narrows disaster, recommended the addition of stiffening girders criss-crossed between the chords underneath the entire roadway. The plan was accepted. In the early 1970s, all the suspender ropes were replaced after corrosion was discovered near the gusset plates at the junction of some ropes and floor-system chords. This took nearly four years to complete.

Work on the roadway was next. By the bridge's fortieth anniversary in 1977, more than 500 million vehicles had crossed it, and cracks were beginning to appear in the road. Fog and salt air sifted into cracks in the concrete and deposited salt chloride into its grain, reaching 3 pounds per cubic yard. Residue from

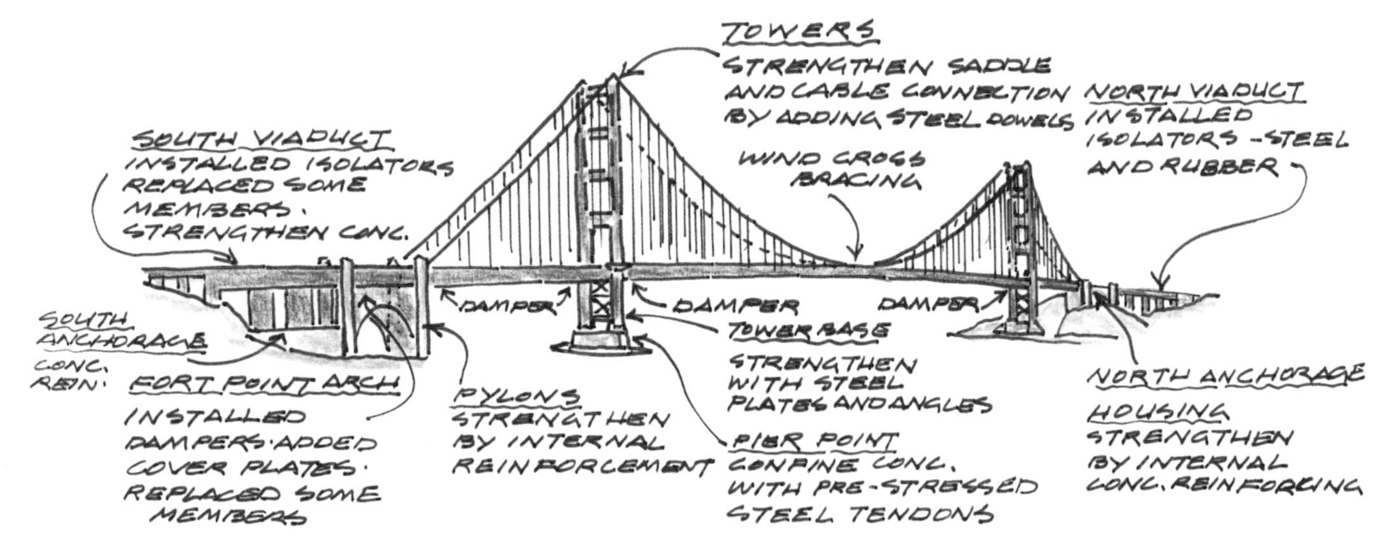

AREAS OF SEISMIC RETROFIT
FOR THE GOLDEN GATE BRIDGE

Opposite: Figure 64

chloride and rust actually had increased the roadway's weight and height, and the chloride was eating into the reinforcing steel below. The only solution was to replace the road. In August 1985, after 401 days and 52.5 million dollars, the job was done—and the curb lanes widened a foot to accommodate the tremendous increase in both traffic and vehicle size. The new, stronger road was also lighter by some 11,000 tons. Yet the road is constantly inspected and kept up: the girders remain free of rust and the foundation piers secure.

After the 1989 Loma Prieta earthquake, however, a vulnerability study indicated that more work needed to be done. The bridge was susceptible to severe damage in an earthquake with a Richter magnitude of 7 or greater and an epicenter near the bridge. Likely to suffer the greatest damage would be the southern and northern approach viaducts, which could easily collapse. The tower saddles, which support the main cables where they cross the tower tops, could also move and be damaged. The suspension span could ram the towers, the Fort Point arch could uplift and move, and the south pylons could be extensively damaged.

A 13.3-million-dollar retrofit, intended to strengthen the bridge enough to survive an earthquake up to magnitude 8.3, began, including structural improvements and efforts to maintain the bridge's aesthetic integrity and appearance (Figure 64). It is too expensive to simply replace the bridge: in 1994–95, the cost of doing so was estimated at 1.4 billion dollars. The towers and foundation of their supports were strengthened and steel elements at the south and north viaducts replaced. Perforated steel plates replaced horizontal laced trusses below the roadway and joined vertical supports. Diagonal box beams under the deck were added for further strength. At the Fort Point Arch, load paths were altered to bypass the critical arch rib members, and some laced members were replaced with perforated steel plates (Figure 65). Damping devices and additional expansion joints were added to dissipate earthquake movement energy.

THESE NEW LACE TRUSSES CAN BE VIEWED UNDER THE NORTH VIADUCT

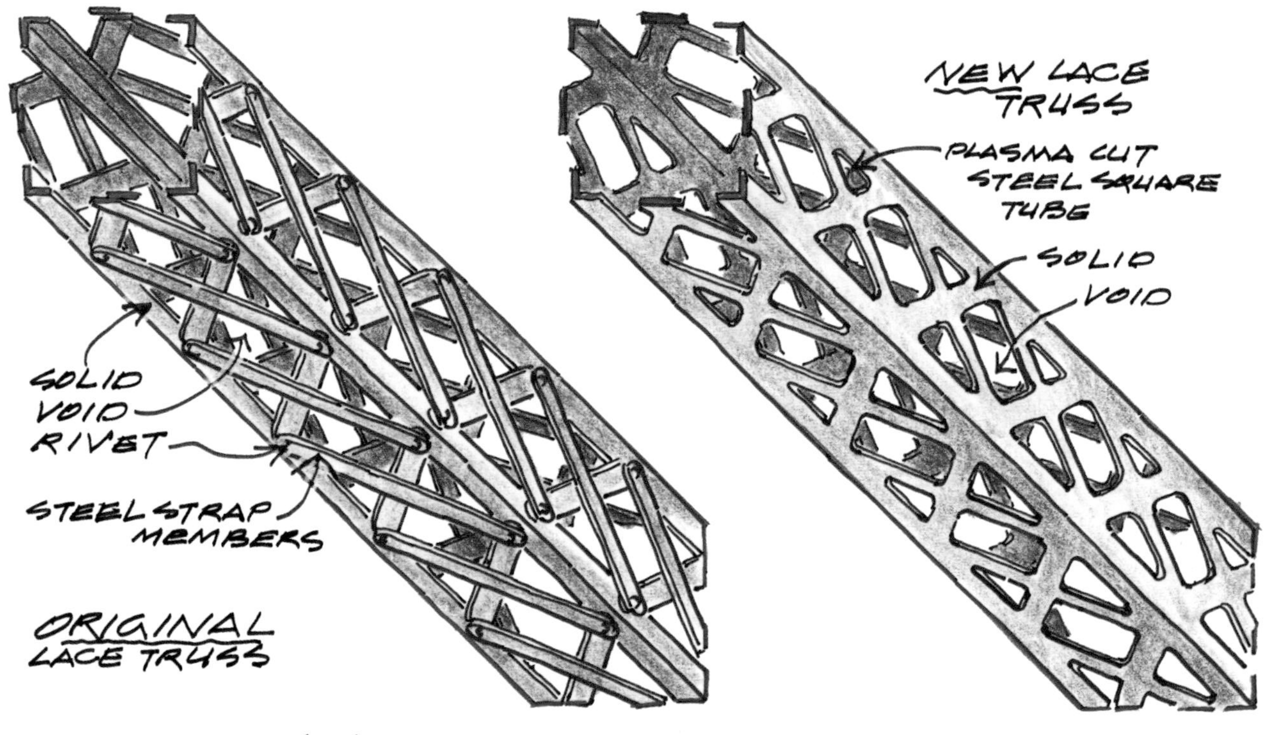

SOLID AND VOID COMPARISON OF THE HISTORIC LACE TRUSS AND THE NEW REPLACEMENT TRUSS

Opposite: Figure 65

The biggest job was on the suspension portion of the bridge itself: its 4,200-foot center span and two 1,125-foot side spans. The two parallel cables—each about 36 inches in diameter, spaced 90 feet apart, with a horizontally projected (that is, when fully straightened out) length of 7,248 feet—needed reinforcement. The 746-foot-tall towers, steel shafts braced with struts, also needed an upgrade. A key remedy was the installation of truss dampers under the deck at the bridge's east and west sides. The deck is designed so that it hangs from the cables and is not attached to the towers. In an earthquake, the truss dampers will keep the deck from hitting the towers with a force that could do serious damage; their installation did not alter the visual appearance of the bridge. The tower shafts, meanwhile, were internally strengthened via the addition of steel diaphragm plates.

Side span reinforcement also strengthened the bridge's stability. Restraints were placed between the steel deck and the existing floor beams, as were saddle reinforcements atop each tower. Strengthened concrete and 20-foot-high steel cover plates, as well as pre-stressed steel tendons, were installed at the south tower base. Similar work occurred at the north tower. To stabilize against wind, the main cable at the center span was attached to the stiffening truss under the bridge deck, and cables and dampers improved aerodynamic stability via the addition of fairings (Figures 66–67). Additional work was done at the north viaduct.

A SUICIDE PREVENTION BARRIER

A controversial issue in bridge retrofitting is suicide prevention. The Golden Gate has the dubious distinction of being the world's number-one suicide site. Since its opening, the Bay Bridge has recorded only 152 suicides; the Golden Gate Bridge, over 1,300. Why? Accessibility is the first reason. Strauss built the bridge to accommodate pedestrians, and the guardrails, which could have prevented jumpers, were instead designed

Figure 66

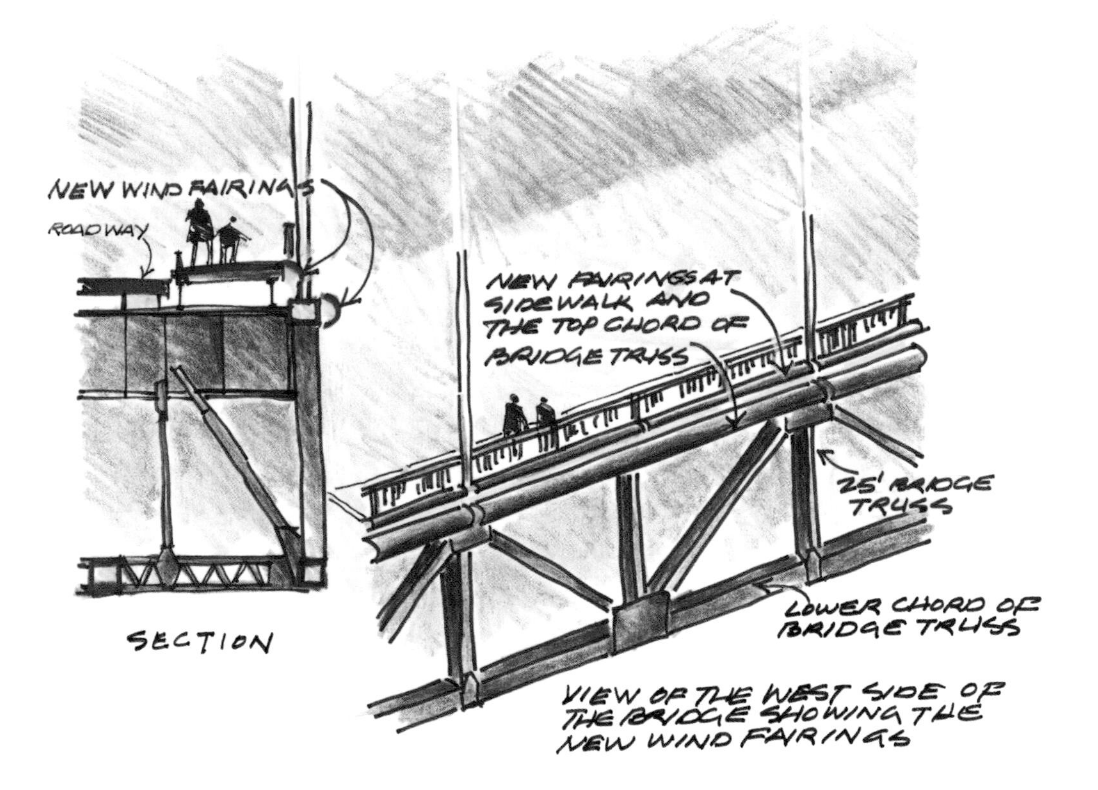
NEW WIND FAIRINGS
ROADWAY
SECTION
NEW FAIRINGS AT SIDEWALK AND THE TOP CHORD OF BRIDGE TRUSS
25' BRIDGE TRUSS
LOWER CHORD OF BRIDGE TRUSS
VIEW OF THE WEST SIDE OF THE BRIDGE SHOWING THE NEW WIND FAIRINGS

Figure 67

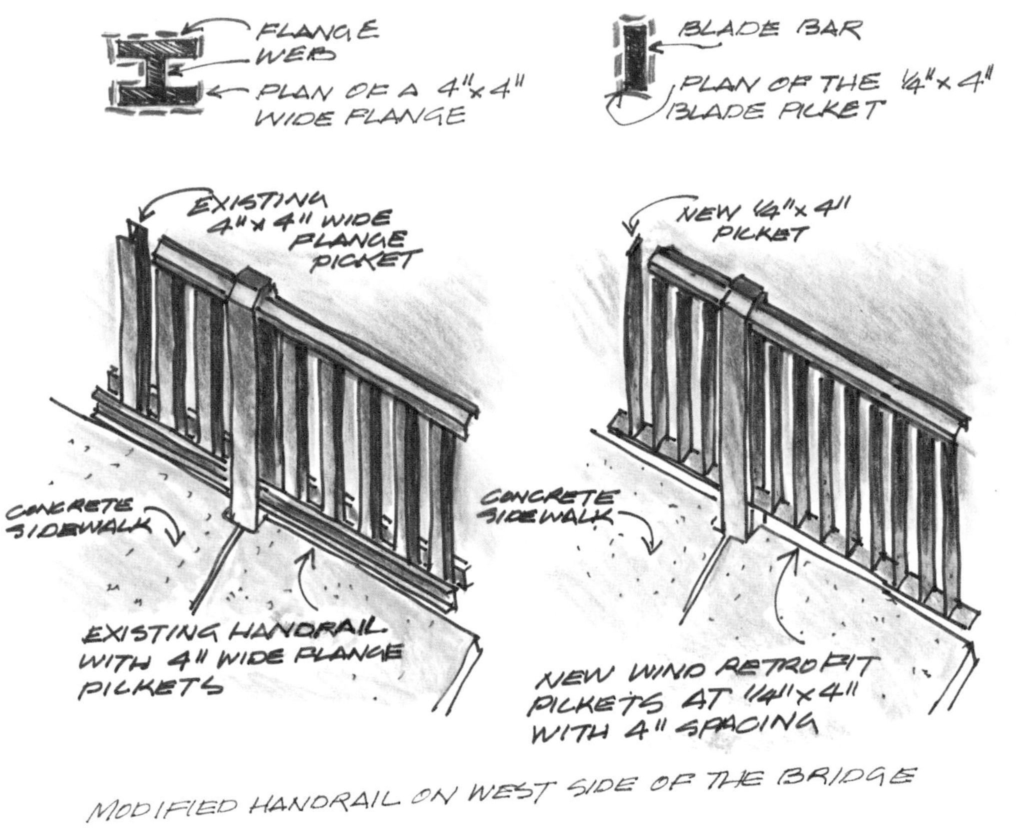
FLANGE
WEB
PLAN OF A 4"x4"
WIDE FLANGE
BLADE BAR
PLAN OF THE 1/4"x 4"
BLADE PICKET
EXISTING
4"x 4" WIDE
FLANGE
PICKET
NEW 1/4"x 4"
PICKET
CONCRETE
SIDEWALK
CONCRETE
SIDEWALK
EXISTING HANDRAIL
WITH 4" WIDE FLANGE
PICKETS
NEW WIND RETROFIT
PICKETS AT 1/4"x 4"
WITH 4" SPACING
MODIFIED HANDRAIL ON WEST SIDE OF THE BRIDGE

Opposite: Figure 68

to improve the view. They are only 4 feet high and easily breached. Most suicides are pedestrians. They favor the east side rather than the west, not so much because they wish to say farewell to the city (or wish to be seen), but because that sidewalk is open seven days a week. The west is open only on weekends.

The bridge had been open less than three months when, on 7 August 1937, Harold Wobber, a World War I veteran, walked to the center span and leapt over the side, falling 260 feet and becoming the bridge's first recorded suicide. Historically, numbers were modest, though possible suicides outnumber actual ones by an average of four to one. Ten people jumped in 1940, and during the war years, numbers declined—two in 1942, three in 1943, four in 1944—but they rose to ten again in 1945. By 1948, the number was nineteen. Throughout the 1950s, suicides averaged ten a year; the figure tripled in the 1960s. In the summer of 1968, a record thirty deaths were recorded. In 1976, forty-two people jumped and one hundred fifty-one were removed for appearing to be suicidal. In 1996, when twenty-one people jumped, seventy were saved. Leaping off the bridge violently and traumatically disfigures the body. One hits the water in approximately four seconds, traveling at 75 miles per hour and landing at 15,000 pounds per square inch. Eighty-five percent suffer broken ribs, which destroy the internal organs. Most who survive the initial fall plunge so deeply—the depth is 350 feet—that they drown. A study of 169 bridge deaths revealed that only 8 died from drowning; the rest died from injuries suffered on impact.

Why do people jump? Psychologists and others who have studied the problem note that jumpers idealize what will happen after they leap. They believe their suicide is "a gateway to another place. They think that life will slow down in those final seconds" and that they will hit the water cleanly, "like a high diver," as writer Tad Friend records. The Golden Gate Bridge, while a symbol of human innovation, may also be a sign of social failure. On average, someone jumps every

VERTICAL STEEL CABLE SYSTEM
EXISTING BIKE HANDRAIL
SIDEWALK
SUICIDE DETERRENT VERTICAL CABLE SYSTEM

Opposite: Figure 69

two weeks, drawn perhaps to the linkage of beauty and death. The bridge's form and grace associate it with a final romantic gesture, but the notes suicides often pin to themselves are unhappy reminders of lives lost: "Absolutely no reason except I have a toothache," one read. Another: "I'm going to walk to the bridge. If one person smiles at me on the way, I will not jump." No one did. There is "a fatal grandeur to the place," Friend suggests, which in 2001 attracted a fourteen-year-old from Santa Rosa, California—not the youngest, however, to die. That was probably a three-year-old thrown over by her despondent father, who quickly followed her down.

Macabre interest in jumpers peaked in 1973, when it was thought the five hundredth victim would soon disappear over the side. Fourteen aspirants were turned back by bridge officials that year. In 1995, number 1,000 approached. To quell interest, the California Highway Patrol halted its official count at 997, although in early July a twenty-five-year-old became the unofficial thousandth victim. He was seen jumping, but like so many his body was never recovered. Among the better-known who jumped are the poet and jazz pianist Weldon Kees, on 18 July 1955, and Roy Raymond, the founder of Victoria's Secret, in 1993. Only twenty-six people have survived the plunge (those who do hit the water feet first and at a slight angle). Only a handful of the ten million or so yearly visitors to the bridge realize that it is a monument to death as well as to the imagination.

For years the issue of a suicide prevention rail has been publicly debated. But unlike at the Empire State Building and the Eiffel Tower, suicide barriers (first proposed in the 1950s) have been opposed at the Golden Gate Bridge. While they unquestionably would reduce the high number of jumpers, suicide barriers, some argue, would also mar the design and the view of the Bay. Cost, aesthetics, and effectiveness were the reasons the bridge oversight board rejected a barrier in the 1970s, although new designs are constantly proposed (Figures 68–73).

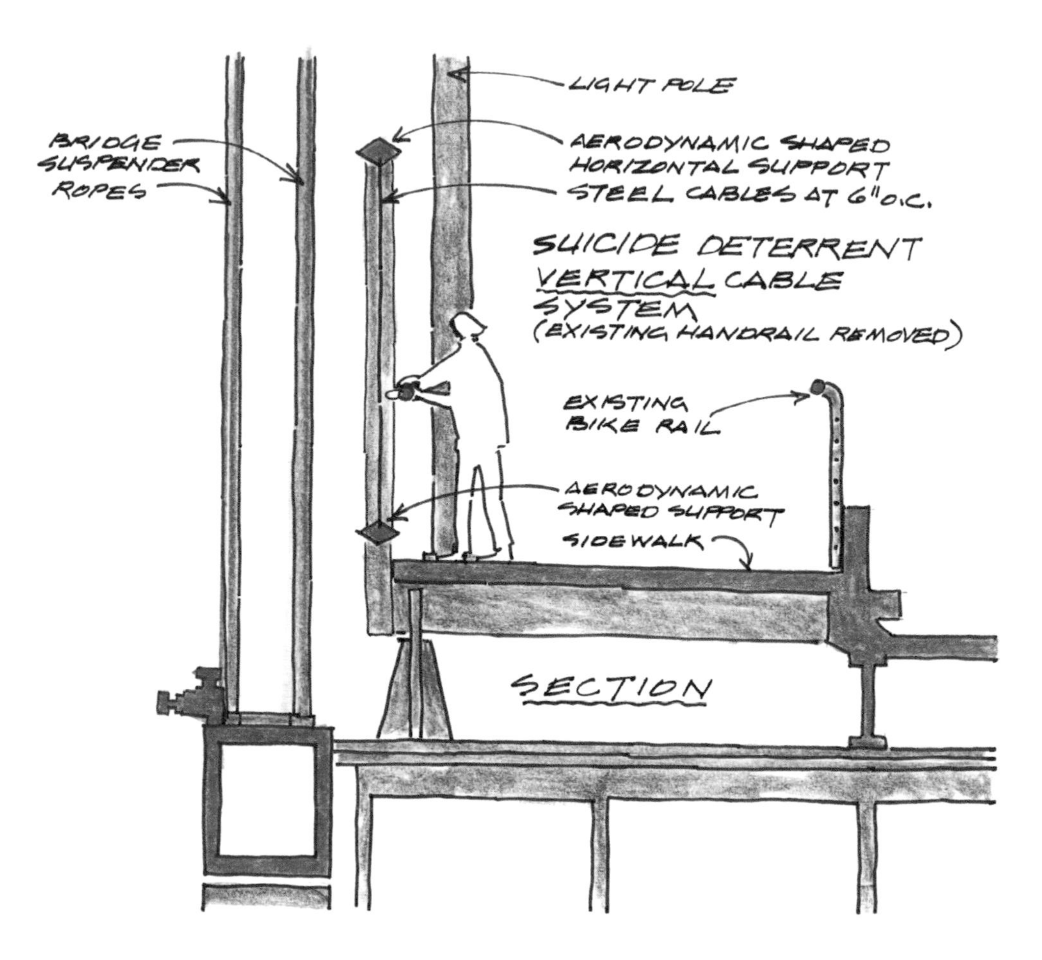
LIGHT POLE
BRIDGE SUSPENDER ROPES
AERODYNAMIC SHAPED HORIZONTAL SUPPORT
STEEL CABLES AT 6" O.C.
SUICIDE DETERRENT VERTICAL CABLE SYSTEM
(EXISTING HANDRAIL REMOVED)
EXISTING BIKE RAIL
AERODYNAMIC SHAPED SUPPORT
SIDEWALK
SECTION

Figure 70

AERODYNAMIC
GLASS WINGLET
HORIZONTAL STEEL
CABLE SYSTEM
EXISTING BIKE
HANDRAIL
SIDEWALK
SUICIDE DETERRENT HORIZONTAL CABLE SYSTEM

Figure 71

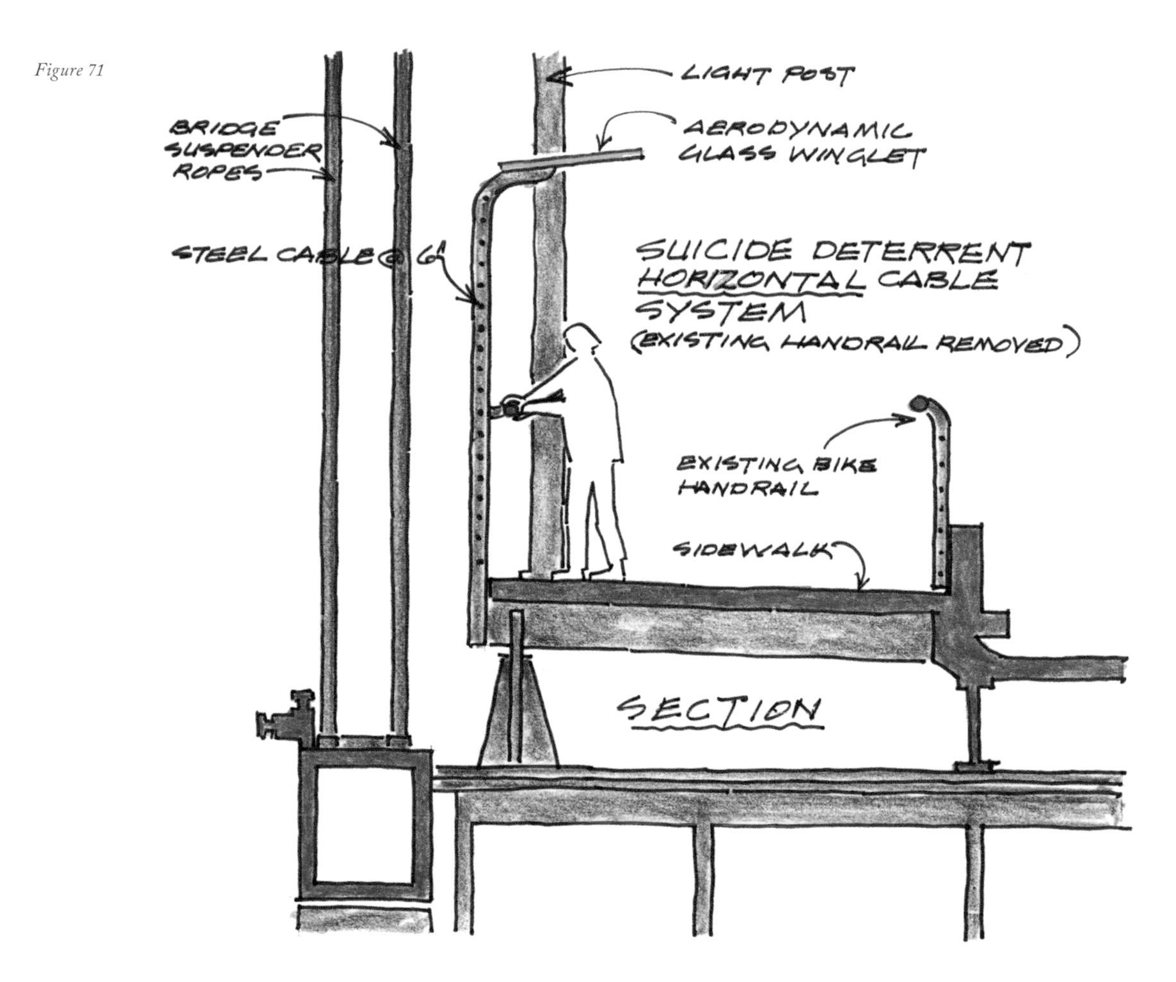
LIGHT POST
BRIDGE SUSPENDER ROPES
AERODYNAMIC GLASS WINGLET
STEEL CABLE @ 6"
SUICIDE DETERRENT HORIZONTAL CABLE SYSTEM
(EXISTING HANDRAIL REMOVED)
EXISTING BIKE HANDRAIL
SIDEWALK
SECTION

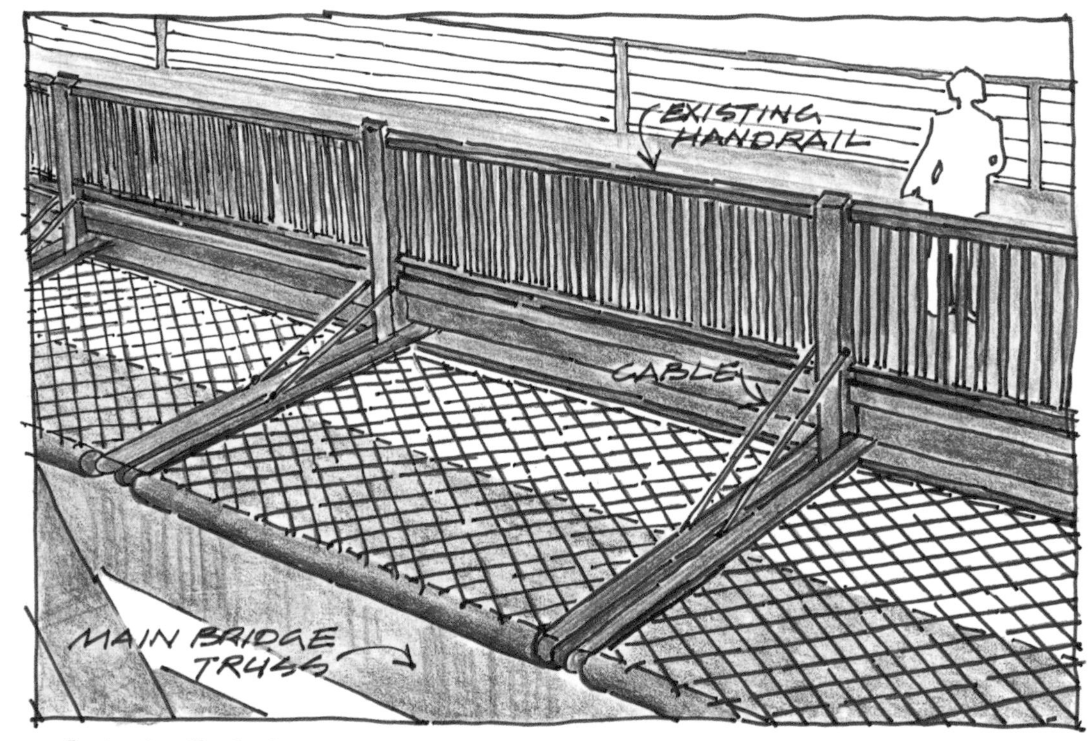

SUICIDE DETERRENT NET SYSTEM

Opposite: Figure 72

Currently a new design study on installation of a suicide barrier is underway. In the meantime, security cameras and thirteen telephones on the bridge act as a warning system, and police and security randomly patrol. This nonphysical barrier currently catches between fifty and eighty people a year but misses approximately thirty.

Despite these tragedies, many share the magical view of the bridge expressed by San Francisco's favorite and best-known poet, Lawrence Ferlinghetti:

At the Golden Gate
A single plover far at sea
wings across the horizon
A single rower almost out of sight
rows his scull into eternity
And I take a buddha crystal in my hand
And begin becoming pure light

Other artists, from photographers to writers to filmmakers—see Vikram Seth's verse novel *The Golden Gate* and Alfred Hitchcock's San Francisco movies, especially *Vertigo*—have found beauty in the bridge. Even Joseph Strauss could not resist the bridge's poetic appeal. For opening day in 1937, he composed "The Mighty Task Is Done," a seven-stanza rhymed poem that begins:

At last the mighty task is done;
Resplendent in the western sun
The Bridge looms mountain high;
Its titan piers grip ocean floor,
Its great steel arms link shore with shore,
Its towers pierce the sky.

Its combination of romance and tragedy makes the Golden Gate Bridge irresistible for suicides and poets alike. An authentic form of art, it also symbolizes a city.

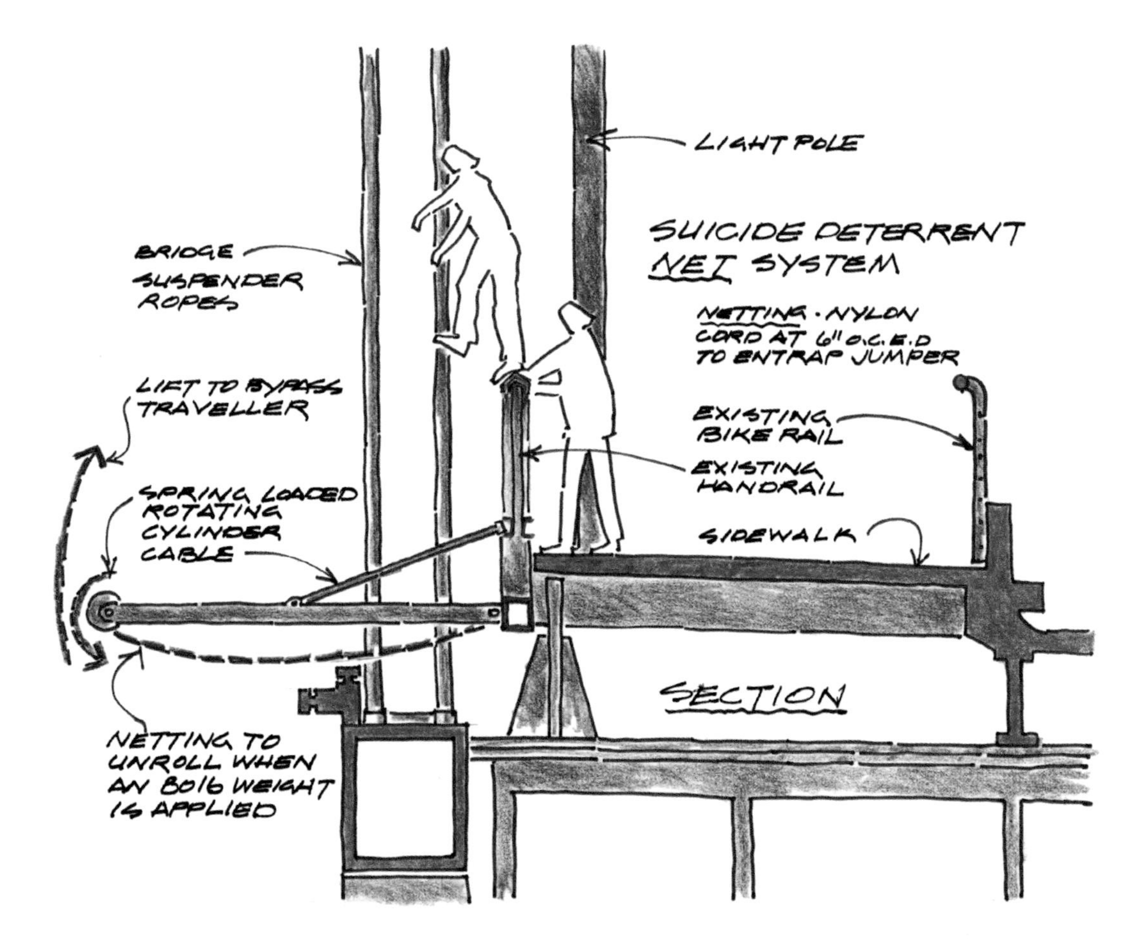
LIGHT POLE
SUICIDE DETERRENT NET SYSTEM
BRIDGE SUSPENDER ROPES
NETTING - NYLON CORD AT 6″ O.C.E.D TO ENTRAP JUMPER
LIFT TO BYPASS TRAVELLER
EXISTING BIKE RAIL
EXISTING HANDRAIL
SPRING LOADED ROTATING CYLINDER CABLE
SIDEWALK
SECTION
NETTING TO UNROLL WHEN AN 80lb WEIGHT IS APPLIED

Opposite: Figure 73

Indeed, San Francisco is the Golden Gate Bridge; which unquestionably speaks for, and to, its people—some of whom come to the bridge to find themselves and leave, while others come to find themselves and, sadly, jump.

The bridge's 1937 opening was as memorable as its subsequent history has been, beginning with the weeklong "Golden Gate Fiesta." Al Jolson sang and fireworks nightly lit celebratory parades, tournaments, and races. Caravans from the eleven western states plus Mexico and Guatemala paraded up Market Street, where the street lamps were decorated to resemble redwood trees. Canadian Mounties and Alaskan dogsleds participated, and a Chilean navy ship sailed into the Bay. A coronation ball was held the night before the opening, ending at 6 P.M. on Thursday, 27 May, when, at last, Pedestrian Day arrived. There were no vehicles, speeches, or ribbon cutting, just thousands of people walking, stopping, and marveling at the structure and the views despite the mist and fog of the chilly day. By 6 P.M., it was estimated that 200,000 people had crossed, including two sisters on roller skates and one man on stilts. The fifty-two-month project had ended in a triumph that continues today.

A DIGEST OF QUANTITIES

BRIDGE FACTS & FIGURES

CHRONOLOGY AND OPERATIONS

Official groundbreaking occurred on 26 February1933, although actual construction had begun on 5 January of that year. The bridge was unofficially opened on 27 May 1937, Pedestrian Day, following a tradition instituted at the Brooklyn Bridge. On 28 May 1937, at noon West Coast time, the bridge was formally opened to vehicles when President Franklin Delano Roosevelt sent a telegraph signal from Washington.

In 1930, Chief Engineer Strauss estimated construction costs at $27,165,000; the final total was $27,125,000 (excluding easement, financing, engineering, and administrative fees, which brought the total to $35,000,000). The bridge has never had a deficit since its opening and continues to pay its way without tax subsidies of any kind. Strauss died almost a year after the bridge opened, on 18 May 1938, at age sixty-eight.

The bridge is operated as a public trust by a board of fourteen directors, responsible for outlining policy, representing the six counties of the Bridge District.

NUMBERS

Each cable weighs 11,000 tons and contains 27,572 separate wires; there are six lanes of traffic on the approach roads. The bridge is 90 feet wide, although the roadway between curbs is 62 feet wide; each sidewalk is 10 feet wide. Its total capacity is 260,000 cars per 24 hours, although this has been exceeded.

The main span is 4,200 feet, or four-fifths of a mile, between the towers, three times the length of the Brooklyn Bridge and 700 feet longer than the George Washington Bridge.

The total length of the bridge proper, including its two approaches, is 8,981 feet, or 1.7 miles; the George Washington Bridge is only 4,660 feet long. The total length of the roadway, from the Marina Gate to Waldo Point in Marin County, is 7 miles.

The towers, 121 feet wide at the bottom and rising 746 feet above the roadway, were at the time of construction the highest and largest bridge towers in the world. Measured from the base of the San Francisco pier, the total height is 884 feet. The cable masts are the tallest in the world. Height of towers above roadway: 500 feet. Dimension of each tower base: 33 by 54 feet. Load on each tower from cables: 61,500 tons. Combined weight of the two towers: 88,800,000 pounds. Approximate number of rivets in each tower: 600,000. Depth of deepest foundation below mean low water: 110 feet.

The total weight of the bridge, anchorages, and north and south approaches, as of 1994, was 887,000 tons. The original weight in 1937 was 894,000 tons.

Minimum vertical clearance at the center is 220 feet above mean high water, 100 feet greater than the Brooklyn Bridge and 20 feet more than the George Washington Bridge. The maximum vertical clearance is 256 feet above mean low water, the greatest navigation clearance in the world at the time of construction.

The two main cables are 36.5 inches in diameter, and each measures 7,650 feet between anchorages, as against 36-inch cables that are 5,270 feet long on the George Washington Bridge. Cable sag, or versine, at the center is 475 feet. At the center of the main span, the cables are only 10 feet above the roadway. Total length of wire used: 80,000 feet. Number of strands in each cable: 61. Weight of main cables, suspender cables, and accessories: 24,500 tons.

ANNUAL VEHICLE TRAFFIC

1937–38: 3,311,512

1938–39: 4,031,504 (22% increase in one year)

1940–41: 4,764,758. Auto ferries across the Bay stopped running in 1940.

1942–43: U.S. Office of Censorship requested that no figures be published for security reasons. Toll-free use by military and federal personnel and their dependents was initiated, but legislation to limit their free travel passed in 1944.

1945–46: First year of published traffic reports stated a daily average of 18,198 vehicles. The state authorized $5 million to build a new six-lane divided highway approach from Marin County to the bridge.

1946–47: On the bridge's tenth anniversary, 7,458,424 vehicles crossed, more than double the figures for 1937–38.

The following years saw increased growth, in part fueled by population growth of 51.6 percent in the north counties between 1940 and 1950. In 1950–51, traffic grew 8.37 percent, and by 1952, the fifteenth year of bridge operation, 11 million vehicles crossed and nearly 46 million dollars in tolls were collected. In 1954–55, 13,220,641 vehicles crossed. So satisfied with the financial state of the bridge was its board that in February 1955, it lowered the automobile toll from forty cents to thirty cents, and in October dropped it again, to twenty-five cents. By 1979–80, it was $1. In the early '90s, a projected shortfall meant increasing the toll to $3; from 1992 to 2002, it remained there, but in 2002 it was raised to the current $5. As early as 1968, the Golden Gate Bridge became the first major bridge to institute one-way toll collection; this soon was imitated worldwide.

Meanwhile, vehicle crossings continued to rise: in 1957–58, the figure was 16,408,399; in twenty years (1977–78), the number had grown to 36,031,236;

and twenty years after that, in 1997–98, the figure climbed to 41,381,800. In 2003–04, however, numbers dipped slightly, to 38,881,684 vehicles. The current average number of daily crossings is 106,525. The highest single-day figure was 162,414 crossings, on 27 October 1989. By 30 June 2004, the combined number of vehicles that had crossed the bridge since its opening reached 1,805,663,417, according to the bridge's Web site (www.goldengatebridge.org).

In 1989, bridge administration explored adding a transit deck for light or heavy rail vehicles, buses, or other mass transportation below the car deck, concluding that the bridge could sustain this addition without significantly altering appearance or structural integrity. The project has remained dormant because of cost.

EARTHQUAKES

Increased use of the bridge meant increased surveillance of its structure and seismic performance. On 22 March 1957 a series of earthquakes shook the area, measuring 5.5 on the Richter scale (the devastating 1906 earthquake was 8.6). Only minor vibrations occurred on the bridge, with vertical movement of just 5.6 inches; the bridge structure, towers, foundations, and anchorages were undamaged. A few windows broke at the toll plaza, however.

The 1989 Loma Prieta quake, while causing limited structural damage, was cause for alarm. On that day, 17 October, all bridge tolls and high-occupancy vehicle lane rules were suspended. Extra buses were scheduled and special arrangements made with trucking fleets that normally do not use the Golden Gate Bridge (the Bay Bridge, which suffered extensive damage, was closed). An all-time high for vehicle crossings over a twenty-four-hour period was recorded: 164,414.

Below: Figure 74

WORLD SUSPENSION BRIDGES

The seven largest suspension bridges in the world by length of center span in 2005

1. Akashi-Kaikyo Bridge, Japan, 1998: 1,991 meters (6,532 feet)
2. Great Belt East Bridge, Denmark, 1997: 1,634 meters (5,328 feet)
3. Humber Bridge, UK, 1981: 1,410 meters (4,626 feet)
4. Jiangyin Yangtze River Bridge, China, 1999: 1,385 meters (4,544 feet)
5. Tsing Ma Bridge, China, 1997: 1,377 meters (4,518 feet)
6. Verrazano Narrows Bridge, USA, 1964: 1,298 meters (4,260 feet; Figure 74)
7. Golden Gate Bridge, USA, 1937: 1,280 meters (4,200 feet)

FURTHER READING

Bennett, David. *The Creation of Bridges.* Edison, NJ: Chartwell Books, 1999. An informed photo book outlining the history of bridges.

Brown, David J. *Bridges.* New York: Macmillan, 1993.

Cassady, Stephen. *Spanning the Gate: The Golden Gate Bridge.* Rev. ed., Santa Rosa, CA: Squarebooks, 1986. A dramatic photodocumentary with an informed text.

Crowe, Michael F. *Deco by the Bay: Art Deco Architecture in the San Francisco Bay Area.* New York: Viking Studio Books, 1995. Excellent photos accompanied by nine walking tours.

Ferlinghetti, Lawrence. "At the Golden Gate." *San Francisco Poems.* San Francisco: City Lights Foundation, 2001, 75.

——. "Inaugural Address." *San Francisco Poems.* San Francisco: City Lights Foundation, 2001, 9–22.

Friend, Tad. "Jumpers." *The New Yorker,* 13 October 2003: 48–59.

Gimsing, Niels J. *Cable Supported Bridges: Conception and Design.* New York: Wiley, 1983.

Grossman, Elizabeth Greenwell. *The Civic Architecture of Paul Cret.* Cambridge: Cambridge University Press, 1996.

Horton, Tim, and Baron Wolman. *Superspan: The Golden Gate Bridge.* Santa Rosa, CA: Squarebooks, 1997.

Kraft, Jeff, and Aaron Leventhal. *Footsteps in the Fog: Alfred Hitchcock's San Francisco.* Santa Monica, CA: Santa Monica Press, 2002.

Morrow, Irving F. "Report on Color and Lighting for the Golden Gate Bridge." Given to the Board of Directors of the Golden Gate Bridge Highway and Transportation District. San Francisco, 1935.

Petroski, Henry. *Engineers of Dreams: Great Bridge Builders and the Spanning of America.* New York: Knopf, 1995. (See especially pp. 272–85.)

Plowden, David. *Bridges: The Spans of North America.* New York: Viking Press, 1974. (For the Golden Gate Bridge, see pp. 249–54.)

Riesenberg, Felix, Jr. *Golden Gate: The Story of San Francisco Harbor.* New York: Tudor Publishing Co., 1940. A readable journalistic history of the harbor and its shipping.

Schock, James W. *The Bridge: A Celebration: The Golden Gate Bridge at Sixty.* Mill Valley, CA: Golden Gate International, Ltd., 1997. A documentary account of the bridge reprinting original sources, from newspaper stories to annual reports and other material, from the archives of the Golden Gate Bridge.

Strauss, Joseph. "The Golden Gate Bridge: Report of the Chief Engineer." San Francisco: Golden Gate Bridge and Highway District, 1938. The most comprehensive and detailed contemporary report; covers the bridge's history, planning, construction, materials, and fabrication and includes remarkable photographs and fold-out engineering drawings.

van der Zee, John. *The Gate: The True Story of the Design and Construction of the Golden Gate Bridge.* New York: Simon and Schuster, 1986. A detailed account emphasizing the politics and history of the construction process.

Wilson, William H. *The City Beautiful Movement.* Baltimore: Johns Hopkins University Press, 1989.

INDEX

G

H

I

J

K

L

M